T0173341

Why Don't Things Fall Up?

Why Don't Things Fall Up?

*and Six Other Science Lessons
You Missed at School*

ALOM SHAHA

HODDER &
STOUGHTON

First published in Great Britain in 2023 by Hodder & Stoughton
An Hachette UK company

2

Copyright © Alom Shaha 2023

A CIP catalogue record for this title is available from the British Library

Hardback ISBN 978 1 529 34816 3
ebook ISBN 978 1 529 34817 0

Typeset in Plantin Light by Manipal Technologies Limited

Printed and bound in Great Britain by Clays Ltd, Elcograf S.p.A.

Hodder & Stoughton policy is to use papers that are natural, renewable
and recyclable products and made from wood grown in sustainable forests.
The logging and manufacturing processes are expected to conform to the
environmental regulations of the country of origin.

Hodder & Stoughton Ltd
Carmelite House
50 Victoria Embankment
London EC4Y 0DZ

www.hodder.co.uk

For Kate, Renu and Mina.

*Those who dwell, as scientists or laymen, among the beauties
and mysteries of the earth are never alone or weary of life.*

Rachel Carson, *The Sense of Wonder*, 1965

The universe is made of stories, not of atoms.

Muriel Rukeyser, *The Speed of Darkness*, 1968

*Practise any art, music, singing,
dancing, acting, drawing, painting,
sculpting, poetry, fiction, essays, reportage,
no matter how well or badly,
not to get money and fame, but to experience becoming,
to find out what's inside you, to make your soul grow.*

Kurt Vonnegut,
Letter to students of Xavier High School, 2006

Contents

Introduction
What's the Point of This?

'Why?' seems to be my daughter's favourite word at the moment. She's just turned three and, like most children around this age, is capable of following up any answer I give her with another 'Why?' and then another and another. This incessant questioning can sometimes be exasperating, but I usually enjoy seeing how long I can go on responding to her before I run out of answers. It probably helps that I'm a science teacher and I've found that she listens quite intently if I talk about *photosynthesis* when she asks about leaves, or the *water cycle* when she asks about clouds. No matter how detailed my answer, though, she always seems to have a follow-up question.

At this point, if I was looking to score easy points with you, I might have written that 'children are born scientists'. This is something that sounds profound, and maybe even makes your spine tingle a little, like the best aphorisms do. But it's just not true. I'd go as far as to say that it's an insidious idea because it suggests *being scientific* is an ability we somehow lose as we grow up, instead of one that requires very specific nurturing to develop. I have taught hundreds and hundreds of children, many of whom have grown up to become scientists, but I can confidently assert that none of them were born with the knowledge, or particular set of intellectual and practical skills, which the role requires.

Young children are naturally curious, and have an innate desire to learn more about the world around them. The insistent 'Why? Why? Why?' that most parents of toddlers are familiar with, is their way of expressing this. But, let me stress again, they are *not* born scientists. Just as anyone who might have an aptitude for art or music or writing needs some form of teaching, and practice, to become really good at these things, children with a natural urge to understand how the world works need to develop, and practise, a distinct set of skills.

Another mistake is to think that science is simply a refined form of 'common sense'. In fact, to understand scientific explanations for natural phenomena, we often need to overcome our intuitive ways of thinking about the world. But what is certainly true is that, without questions, there would be no science. At its heart, science is a process by which we find answers.

This is a book for anyone who is curious. It's a book for anyone who didn't understand science at school, anyone who simply didn't like their science lessons, or was left bored or confused by them. This is a book for anyone who, when faced with a question about the world around them, has wished they could remember more of what they should have learned.

Most 'popular science' books, which attempt to explain scientific ideas, are written by scientists or former scientists. In my experience, these books are often written with an assumption that the reader already has some basic knowledge, and is interested and motivated enough to want to deepen it. I'm a secondary school science teacher, and I'm bringing the perspective of someone who has spent much of the past twenty-five years trying to explain science to people who don't understand it at all, and sometimes have little inclination to try. I have a lot of empathy for such people because, until the later years of my secondary education, I was one of them.

The questions in this book are ones that many young children ask, and are deceptively simple. Like fast food, fast explanations to questions such as 'Why is the sky blue?' and 'What am I made of?' are rarely satisfying; they often rely on making dangerous assumptions about what people already know, and can lead to confusion and misunderstandings later on. In giving *my* answers to the seven questions in this book, I hope to take you on a journey through the 'big ideas' of science, which any decent school science education should cover.

Along the way, I'll remind you about the science of everything from atoms to cells to stars, as well as showing you the power and beauty of scientific achievements like the Big Bang Theory and evolution by natural selection. I've used explanations that I've honed over many years of teaching, taking into account the most common misconceptions and intellectual hurdles that prevent people from grasping key ideas. I've also included descriptions of some of the experiments and practical experiences that a good science education ought to include, as well as some of the stories which best illustrate how scientific discoveries are really made, as opposed to simplistic narratives of lone geniuses and single moments of inspiration.

The most important question I have to answer as a teacher is: 'What's the point of this?' I could point out that the society we live in is built on the fruits of science – from the electricity that powers our lives, to the technologies that keep us connected and entertained, to the medicine that allows us to survive illness and disease and live longer than ever before. I could also tell you that it's important to understand science so that you can make sense of issues like climate change and stem cell research, and make well-informed decisions when these matters become the focus of political decisions for which you have a vote.

I could tell you that science plays a crucial role in helping us deal with many of the problems we face as a species, from preserving the environment to ensuring there is enough food for everyone to eat. While these are all important reasons why science is given such a prominent place in education systems around the world, they are not really why I took an interest in science myself, or why I think everyone should have the opportunity to learn about it.

For me, there are two main reasons why everyone should be entitled to a good science education. First, it is a uniquely powerful 'intellectual toolkit' for helping us to make sense of the natural world. Second, and perhaps more importantly, it is a *cultural* activity with which all of us should be able to engage, and indeed contribute to, should we wish. I think science comes from the same thing that drives us to make art, music and literature – our response to being alive, existing in this world, and the urge to make sense of it and share what we have understood with others. Science is not only for those who want to become scientists, just as art is not only for those who wish to become artists; it is a *human endeavour* that enriches all our lives.

Scientific ideas and discoveries have the capacity to elicit awe and wonder from us, and I hope you will experience that in the pages of this book. But I think the real joy of science is to be had from an understanding of it that is genuinely meaningful and useful. For that, you need to start with the basics, simple models that you can get your head around and apply to the world around you. That's what I'm hoping to do with this book – to introduce you (or remind you about) the scientific ideas you *should* have learned at school, so that you can provide truly satisfying answers to questions like 'Why don't things fall up?' and also have the confidence to be unashamed about saying, 'I don't know, that's an interesting question, how can we find out?'

This book is by no means a definitive guide to science, not even to the science that is covered in schools. Instead, it's a journey through some of the key ideas, which I hope will give you a taste for more, in the same way that listening to a song you love for the first time might give you the urge to listen to more songs by the same artist, or reading a great book makes you want to read all of that author's work. This is my attempt to write a popular science book for people for whom science was *not* popular, for anyone who wants to give science a second chance to make sense.

Chapter 1
Why Is the Sky Blue?

Trying to answer the question 'Why is the sky blue?' takes me to the heart of what this book is about. I believe that a brief answer can only, at best, provide an *illusion* of understanding unless the person asking the question has a fairly good grasp of the physics they should have learned at school. What I hope to do is give *you* enough understanding of the science to start using this wonderful and powerful way of making sense of the world for yourself.

Let me show you what I mean – here is the short answer provided on a website for parents and educators: 'Sunlight reaches Earth's atmosphere and is scattered in all directions by all the gases and particles in the air. Blue light is scattered more than the other colours because it travels as shorter, smaller waves. This is why we see a blue sky most of the time.'

There is nothing incorrect about this statement, but if I didn't know any physics, I might wonder what it meant for sunlight to be 'scattered' and how the particles in the air (if I even knew what 'particles' were and that they existed in the air) made this happen. I wouldn't necessarily know that there are different colours of light in sunlight, and I doubt I'd understand what 'travels as shorter, smaller waves' meant or why that was relevant. The website authors do go on to try and explain some of these things, but my point is that an answer can be correct without being much use to the person asking the question.

Here's another short answer (this time from someone on Twitter, where people are limited to using just 280 characters): 'Light is made of lots of different colours. Which is why we get rainbows. The light bounces around in the sky and blue bounces around more than the other colours making the sky look blue.' I rather like this answer – I think it's better than the previous one because it doesn't use technical terms or introduce unfamiliar ideas like scattering and waves. But since I have more than 280 characters to play with, I'd like to give you an answer that, I hope, will give you a deeper level of understanding and be ultimately more satisfying.

I think a good place to start is by pointing out something that may be obvious – the sky isn't always blue, it's pretty much black at night-time, it can be many shades of red at sunset, and for long periods of the year in England it's grey during the daytime. To understand why the sky can be all these different colours we need to know a little about *light*, and how it allows us to see things.

Light is a *thing*. It turns out to be not quite like any other thing, which has made it a very difficult thing for scientists to understand. You may have been told that light 'is a wave', or perhaps you've come across the idea that it's made of particles called 'photons', or maybe you've even read somewhere about 'wave-particle duality', which says that light is *both* a wave and a particle at the same time. These scientific ideas can all be useful when trying to explain the different things that light *does*, and putting them together to understand what light *is* has led scientists to some mind-boggling conclusions about the nature of space and time. I'll come back to wave-particle duality later in the chapter but first, like we would in school, let's cover the basics.

One of my favourite words in the English language is 'luminous'. Like many parents, I think my children are luminous but, in the strictly scientific sense, the word is used to describe *objects which give off light.* The Sun, candle flames and lightbulbs are all luminous objects – in other words, they are *sources* of light. Things like the Moon, and indeed my children, are not luminous because they do not give off their own light. We can see luminous objects because the light they emit travels into our eyes and makes our brains do whatever they do to produce images in our minds. I like this as a good basic description of light:

> *Light is something that travels out from luminous objects and interacts with any object in its path.*

It can seem that light does not actually travel, but somehow *instantaneously* fills space whenever you turn on a lightbulb or light a match. This is because light travels incredibly fast, at a speed of about 300 million metres per second (or, as I like my students to remember it, three times ten to the 8 metres per second, 3×10^8 m/s). This is the speed at which light moves through empty space, and pretty much the same speed it goes through air. If we could travel this fast, it would take less than two-hundredths of a second to get from London to New York, so we cannot possibly notice the time it takes for light to get from one side of a room to the other when we turn on a lamp.

I am writing this at my desk at home, where there are three sources of light – sunlight through my window, an overhead electric lightbulb and the computer screen in front of me. The recent rain has cleaned my windows and, if it wasn't for some faint reflections of the stuff in my room, you would never know the glass was there – from where I'm sitting, the world outside looks exactly the same as it would without the window.

My desk is cluttered with paper and various items of stationery, including a matt black hole puncher, a small plastic box of shiny silver paperclips, a transparent purple Sellotape dispenser and the glasses I wear for watching TV or driving. None of these things give off light and the only reason I can see them is that light from the Sun, the computer and the lightbulb is bouncing off them and into my eyes. The technical term for this is 'reflect': *we can only see a non-luminous object if light from a luminous object reflects off it and into our eyes.*

If there was no window in my room, and I turned off the computer and ceiling light, all these things would appear black. In fact, if I could stop any light at all from getting into the room, I wouldn't even be able to tell that there were different objects in the room, because all I would see is black. The items on my desk all *look* different because light *interacts* differently with each one: *light can pass through an object, bounce off it, be absorbed by it, or some combination of those three things.*

The scientific explanation of why the sky is blue is connected to the explanation of why my hole puncher is black, why my Sellotape dispenser is purple and see-through and why my glasses allow me to see distant things in focus. To understand and explain things like this using *science*, scientists come up with *scientific models*.

Models

In everyday life, the word 'model' might make you think of a tall, good-looking person wearing designer clothes (so, the literal opposite of me). Another common use for the word is to describe a small-scale replica of an object such as a toy car, aircraft or railway. A real railway and a model railway are obviously not the same thing, but they have many of the same features and a good model can show *accurately* how the parts of the real thing fit together and how it works. This is perhaps closer to the meaning it has in science – a scientific model is a way of *representing* something to help us understand it.

Scientific models are usually *simplifications* and *representational* – they describe and explain the behaviour of something by comparing it to things we already know and understand; for example, we can model the behaviour of solids, liquids and gases by representing them as being made up of tiny balls. Scientific models do not usually explain *everything* about the thing they describe, but one of the reasons why they are useful is that they can *predict* how something will behave. When the thing we are trying to understand or explain follows mathematical rules, we can use calculations to make predictions about its behaviour, so scientific models are often also mathematical models.

The nature of scientific models is something that I think is often glossed over, or even completely missed out in people's science education, and this can be a hindrance or even a fatal barrier to understanding science. I remember struggling when we studied electricity in my physics lessons at school – I found everything about the topic confusing, and could not understand how what we were learning in class was supposed to help us *understand* electricity.

One of the things I had started to love about my science lessons was that I felt they were giving me this incredible insight into how the world worked, how it actually *was*. I had particularly enjoyed learning about Newton's laws and the elegant way they described how and why things move. But with electricity, this didn't seem to be the case – what we were learning in class didn't seem to hit that same sweet spot of 'understanding'. I wanted to know *exactly* what electrons *were* and *why* they did what they did in electrical circuits, but all we seemed to be doing was drawing diagrams and trying to work out which lightbulbs would be bright and which would be dim.

I can't remember the exact details, but a conversation with my physics teacher, Mr York, about why I was finding it difficult was a real turning point for me – he was the first science teacher to explain to me that it would take time for me to achieve a satisfactory 'understanding' of electricity, that what I was learning at that point in my school career was a *model*, a set of rules that electrons in electrical circuits followed, and there were other models, rules and ideas about electrons that I would learn later, which would develop my understanding. But, he pointed out, we cannot know what an electron is in the same way we might know what a tree or football is – all we have is the model.

Mr York was the first teacher to make me realise that 'science' is not made up of 'ultimate truths' about how the world is, but of *models* that describe, explain and predict the behaviour of things in the natural world.

The Ray Model of Light

The first model of light that we learn about at school, and indeed one of the earliest that humans seem to have come up with, is the *ray model of light*, in which we use *imaginary* lines to show us the path that light takes. Using rays to represent light, we can explain many of the things that light does – shadows are formed when an object blocks rays of light, we can't see round corners because rays of light travel in straight lines, lenses work by changing the direction the rays are travelling and mirrors work because of the way rays reflect off smooth surfaces.

The ray model of light allows us to explain why objects look the way they do. When rays of light hit an object, they can be absorbed (meaning they do not pass through), or reflected (meaning they bounce off and change direction), or they can be partially absorbed and partially reflected. Light rays can also change direction when they pass through something, an effect called *refraction*, which can be used to explain how, for example, a magnifying glass works and why diamonds sparkle.

One of my favourite activities to do with young students when teaching about light, and the ray model, is to use a pinhole camera. This is simply a small black cardboard box or tube that has a 'screen' made of tracing paper at one end and a tiny hole made with a pin at the other. If it's a bright, sunny day, I turn off the classroom lights, partially lower the blinds and get the students to point the pinhole end of their 'cameras' at the window and look at what they see on the tracing paper screen. If you have not done this for yourself, you may not understand the sheer delight that students invariably experience when they see the view through the window replicated on their screens. (If you've never used a pinhole camera, it's very easy to make one and absolutely worth doing – just use the instructions at the back of this book.)

After the initial 'ooh, look at that', they usually notice something else unexpected: the image is upside down. This can be explained using the ray model of light – imagine the camera is pointed at a tree, the image is formed by rays of light from the Sun bouncing off the tree and into the camera through the pinhole. If we drew a diagram of the rays entering the pinhole and hitting the screen at the back of the camera, we would see that light rays from the top of the tree move in a downward path and hit the bottom of the screen, and rays from the bottom of the tree move upwards and end up on the top of the screen, so the tree appears upside down.

Can you think how the ray model of light can explain why, if we make the pinhole bigger, the image becomes brighter and blurrier? The answer, using the model, is that with a tiny hole, each point on the screen receives a ray of light from only one point on the tree, which gives a clear image. With a bigger hole, a point on the screen receives rays from several points on the tree, making it brighter, but messing up the image. Our eyes work in a similar way – light travels through the small hole in the front of our eyes (the pupil) and the image formed on the back of the eyes (the retina) is upside down, but our brains flip it so we see the world the right way up.

There are lots of fun activities to do with light in school and, despite the challenges of trying to manage thirty children in a darkened room, I like to do some practical work so students can see for themselves that rays of light follow certain rules when they bounce off or pass through things. If you remember any of these rules from school, it's probably 'the law of reflection', which says that when a ray of light reflects from a surface, *the angle of incidence is equal to the angle of reflection*. This means that the angle an incoming ray makes with the surface is the same as the angle the ray makes when it bounces off, a bit like the way a football bounces off a wall (if it isn't given any 'spin'). If an object has a very smooth, flat surface, light rays coming from a particular direction all bounce off in the same direction. This is called *specular reflection*, and is the reason why we see reflected images in mirrors and other objects with very smooth, flat surfaces.

Light is also reflected off objects with rough surfaces (otherwise we wouldn't see them), and each ray of light still obeys the law of reflection at the point where it hits the object. However, different parts of the surface of a rough object point in different directions, so light rays coming from one direction are reflected in different directions. This is called *diffuse reflection*, and explains why we see no image (or sometimes a distorted image) reflected from rough surfaces.

There is also a 'law of refraction', which can be used to calculate the direction rays of light will bend when they pass through a transparent object. When I was at school in the 1980s and early 90s, I was taught this as 'Snell's Law', named after the Dutch scientist Willebrord Snellius, who lived in the 1600s. Long after leaving school, I learned that the same law was independently arrived at by both Thomas Harriot and

René Descartes in the 1600s, and that it had been written about more than five hundred years before that, in the 900s, by the Persian scientist Ibn Sahl.

What's interesting about this is not just that my old school textbooks didn't feature any non-European scientists, but that it also illustrates something fundamental about science: unlike the works of Shakespeare, for example, which could only have been written by that particular person, scientific laws are not 'authored' by scientists, but rather *discovered.*

Using the laws of reflection and refraction, it's possible to draw *ray diagrams* that we can use to *predict* the behaviour of curved mirrors and lenses before we make them, which is useful if you want to build telescopes or cameras or spectacles. I have to confess that I found drawing ray diagrams in my physics lessons rather tedious – you need to use a ruler and protractor to get the lines precisely straight and at the correct angles, and they are time-consuming to draw correctly. With my own students, I like to show how the ray diagrams really do represent the behaviour of light by using a special set of apparatus that lets me show them the path of a laser beam as it travels through lenses and bounces off mirrors.

In movies, laser beams are often shown as straight, glowing lines (usually being fired from weapons), but if you've ever seen a laser pointer in use, you'll know that you can't actually see its beam in the air, you only see the spot where it hits a surface and bounces off. It's very easy to impress a room full of students by turning off the lights and spraying a fine mist of water in the path of a laser beam – it suddenly becomes visible, just like in the movies.

The reason why you can see the laser beam when it's travelling through the mist is that rays of light from the laser bounce off the tiny droplets of water in such a way that, instead of travelling straight on, some of them travel in the direction of your eyes. This effect, where light hits small particles and is bounced away from its straight line path, is called *scattering* and is central to understanding why the sky is blue.

Most people think that clean air is invisible, that we can see right through it because light simply goes through the air without being affected. So, what does the ray model, in which light rays travel in straight lines, say the sky should look like if light does *not* interact with air? It should look black, except for the Sun, which should appear as a bright disk because its rays should travel *directly* into our eyes. Does that mean the ray model is wrong? Not necessarily – we can conclude from the fact that our sky is not black that the rays of light *must be affected* in some way when they interact with the atmosphere. We can also see that the ray model correctly predicts what the sky should look like from the fact that the sky on the Moon, where there is no air, is black.

The reason why we don't see a black sky during daytime is because sunlight is *scattered* by the 'stuff' that makes up what we call air. If we ignore clouds and things like smoke and other pollutants, air consists mostly of molecules of gas. We can't see these particles because they are too small, but when light hits them, it bounces off and spreads out away from its straight line path. It is this *scattering* of light that allows us to see the sky. But this doesn't answer why the sky is blue when the light from the Sun appears to be white – the answer to this is related to something else we see in the sky: *rainbows*.

Rainbows

No matter how many times I see them, I always experience a thrill when there's a rainbow in the sky. They are a true wonder of the natural world, and seem magical even when you know how they're formed. It's not surprising to find rainbows in myths and storytelling from around the world, representing a bridge between our world and another, a bow to shoot arrows of lightning and a sign of a god's promise not to flood the world again.

If you want to see one, you need to have the Sun shining brightly from behind you and water droplets in front of you. You can wait for a rainy day where the circumstances are just right or, as many people discover to their delight, you can make your own rainbow on a sunny, rainless day by using a hose pipe to spray a mist in front of you with the Sun at your back.

I love doing this with my two daughters and hearing their shrieks of joy when they spot the rainbow right in front of them. Sometimes one sees a rainbow before the other, because they're not actually seeing the same rainbow. In fact, none of us can ever see the same rainbow as another person. This is because a rainbow is not something out there in the world in front of us, but something that only exists, in a sense, in our eyes. A rainbow is what we see when sunlight has bounced through raindrops on its way into our eyes.

One of the few bits of scientific knowledge from school that seems to stick with a lot of people is the fact that white light is made up of the colours of the rainbow. I'm not sure why this is, but I think it's got something to do with our fascination with this beautiful phenomenon and finding out that it has a deep connection to something as fundamental as the nature of light. Perhaps lots of people remember this fact because it's quite surprising when they first learn about it; after all, it's by no means obvious and it contradicts our childhood experiences of mixing all our paints together and getting black. Or maybe it's just easy to remember. Whatever the reason, knowing that white light is a mixture of colours leads us back to our explanation of why the sky is blue.

I explained above that the reason why we can see the sky at all is because light from the Sun, which is white, is scattered as it passes through the air. The reason why we see a *blue* sky is that when white light from the Sun, which is a mixture of colours, collides with particles in the air, the blue rays are scattered more than other coloured rays, so when we look up at the sky, away from the Sun, it's mostly blue light that is travelling into our eyes.

This can also help to explain why the sky can appear red at sunset: as the Sun is lower in the sky at this time, its light rays travel a longer distance through air, so that by the time light from the Sun reaches our eyes a lot of the blue light in it has been scattered away and the remaining light is mostly red. When the Sun has sunk below the horizon, the last colour we see in the sky is the deep indigo or violet colour at the end of the spectrum.

Knowing that white light is made up of different colours lets us understand a bit more about how we see coloured objects in general. A simple model tells us that when light hits an object, some colours are absorbed, and some are reflected. So, a blue object looks blue because it only reflects blue light and absorbs all the other colours of the spectrum. Using this model, can you work out what colour a blue object would look like if you shone red light on it? The answer is it would look black, because it would absorb the red light and not reflect anything back. Objects that look black in white light *absorb* all the colours equally, and objects that look white *reflect* all the colours equally.

One of the first songs my children learned was 'I Can Sing a Rainbow', the first verse of which is:

> Red and yellow and pink and green
> Purple and orange and blue,
> I can sing a rainbow,
> Sing a rainbow,
> Sing a rainbow too.

Whilst this is a charming song with a delightful tune, it teaches children an incorrect order of the colours in a rainbow. From top to bottom, the colours in a rainbow are actually red, orange, yellow, green, blue, indigo and violet. Unfortunately, this order doesn't work very well when sung to the tune of 'I can sing a rainbow', but you can remember it by learning the name Roy G. Biv or a mnemonic along the lines of Richard Of York Gave Battle In Vain. It's likely that your science teacher showed you a beam of white light being split up into these colours by shining it through a triangular piece of glass, a prism.

You may also have been told that this is what Isaac Newton did to prove that white light was in fact made up of different colours. In my experience, there's a subtle part of this story that teachers either miss out when teaching it, or simply forget: Newton was not the first person to see that different colours of light emerged when white light is passed through a prism. Lots of people before him knew about this, but none of them came to the conclusion that this meant white light was made up of different colours of light. Instead, they believed that white light was *pure* and that it became coloured when it passed through a prism, because prisms somehow *added* colour to the white light.

Newton tested this idea by doing an experiment: he passed white light through a prism and then he passed the individual colours of light, one by one, through a second prism and found that the coloured light did not change – it came out of the second prism the same colour as when it went in. So the prism was not *adding* anything to the light. Newton also showed that by putting all the colours of light back together, you get back to having white light.

Newton had come up with, and carried out, an experiment that no one before him had thought to do and found out something fundamental about the nature of light. He gave the name 'spectrum' to the colours of light that emerge when white light is shone through a prism. This is only one example of Newton developing scientific ideas that we still learn about and use today, but there's something else they don't usually tell you at school: Newton originally said he saw only five colours in the spectrum, but later changed this to seven, because it matched the number of notes in a musical scale (do, re, mi, fa, sol, la, ti), something he found pleasing.

Not many people can distinguish between indigo and violet, so they might say there are only six colours in the spectrum, and a few people may be able to distinguish more than seven, but Newton's claim that white light is made up of seven distinct colours has stuck.

Although Newton's work was a major step forward in our understanding of light, he still didn't know exactly what light was. Scientists had a better explanation for the colours of a rainbow after Newton's work, but they did not have a model to explain the differences between the different colours of light. There were lots of other things that light did which they couldn't explain, from the colours on the surface of a soap bubble to the 'halo' seen around streetlights on a misty night. To explain these things, scientists needed the *wave* model of light.

Waves

The waves most people are familiar with are those on the surface of water, from the big ones that surfers ride to the small ripples caused by raindrops falling on a puddle. Ripples on water turn out to be a good *model* for both light and sound because, perhaps surprisingly, they have strong similarities in the way they behave. Understanding this has led to technological innovations ranging from microwave ovens to mobile phones and noise-cancelling headphones.

In schools, science teachers often use a piece of apparatus called a 'ripple tank' to demonstrate the behaviour of waves, but I always start the topic of waves using a machine made of . . . jelly babies. The introductory lesson on waves is one I love teaching every year because, from the moment they walk into my lab and see it stretched from one end of the room to the other, my students are always fascinated by the jelly baby wave machine. It looks like some sort of weird, colourful sculpture from a museum of modern art, and nothing like a piece of scientific apparatus. Once they get closer and can see it's made of jelly babies, their curiosity is thoroughly piqued and I definitely have their attention. The best thing about it though? Once I've finished showing them the machine in action, we eat the jelly babies.

I'm going to try my best to explain waves to you in words, but as a teacher I believe it's really important, whenever possible, for my students to *see* the natural phenomena I'm talking about. I think the starting point of science is in *looking really closely at the world*, so I spend a lot of time in my lessons getting students to look at things, and pay attention to what they're seeing in a way they may not usually do.

There are instructions for building your own jelly baby wave machine at the back of this book, so you can see the properties and behaviour of waves for yourself. I really hope you'll try this, but if it's too much trouble, please at least get a large bowl of water (the bigger the better), so you can look closely at the behaviour of waves in water for yourself.

If you briefly dip your finger into some still water, you'll see ripples move outwards, away from your finger. A good model of what is going on can be seen in Mexican waves made by spectators at large music or sports events. The wave clearly moves round the stadium, from one place to another, but the spectators simply stand up and sit back down in their seats (or raise and lower their arms) and end up in the same place they started.

In the case of a Mexican wave, each person knows to stand up after the person next to them stands up, but in a wave on water, each bit of water pulls or pushes a neighbouring bit of water and it's this pulling and pushing that is passed along. Dipping a finger into water causes the water at that point to bob up and down, because the water behaves a bit like a spring – the water at that point is vibrating. These vibrations die down quickly if you only dip your finger in once, but if you repeatedly move your finger in and out of the water, you'll notice a constant stream of waves moving outwards. It turns out that *all waves start with something vibrating.*

A small object, like a leaf, floating on the water, will bob up and down as the waves pass it. Contrary to what many people imagine if they've never looked closely at a situation like this, the leaf is *not* carried outwards, away from its original position. The leaf moves up and down because it gains *energy* from the water, and that energy originally came from you. But it was not the water that you touched with your finger that carried the energy to the leaf, that particular bit of water only moved up and down, *not* horizontally, and once the wave dies away, the water is still in the same place – it is *only the energy that has moved* from one place to another. This is the key thing to notice about this situation, and gets to the heart of what a wave is:

Waves transfer energy from one place to another without transferring any material.

This model allows us to explain how we receive energy from the Sun – it travels to us as light waves. When we see things because of light, something in our eyes is made to bob up and down like the leaf on water.

It is always useful in science to be able to measure things and in the case of waves we can measure two key properties. First, the *frequency* of a wave is the number of waves the source sends out every second and is measured in *hertz* (Hz), where 1 Hz is equal to one wave per second. Second, the *wavelength* is the distance between the peak of two consecutive waves and is measured in metres (m). The wavelength and frequency are connected to each other – if you *increase* the frequency of your ripples by dipping your finger in and out more quickly, you'll notice that the wavelength *decreases,* because the waves are being produced closer together.

If you look closely though, you would notice that the *speed* at which an individual wave travelled across the water would stay the same. Imagine you had a giant tank of water and some kind of machine that was producing waves at a frequency of 3 Hz (that's three waves per second) with a wavelength of 2 metres. Can you see that after one second, three waves would have been produced and they would cover a total distance of 6 metres (three times two) away from the wave source? The first wave would be 6 metres away, so it would have travelled 6 metres in one second – in other words, the wave must have travelled at a speed of 6 metres per second.

If that makes sense to you, hopefully you can see that the speed of the wave is equal to the frequency multiplied by the wavelength – so we have a *mathematical relationship* between the speed, frequency and wave-length of a wave:

$$speed = frequency \times wavelength$$

I'm trying my best to keep the maths in this book to a minimum, but I think it's useful to see how we can arrive at mathematical relationships that become part of a model we use in science. In this case, this equation means that if we know any two of the speed, frequency or wavelength of a wave, we can always calculate the third. This is an example of how a scientific model can be used to *predict* the behaviour of something – the equation above lets you predict the speed of a wave if you know its frequency and wavelength.

Experimenting with a bowl of water can help us understand the properties of waves. And there's more about waves you can learn with such simple apparatus. For example, if you look closely at what happens when the waves reach the end of the jelly baby wave machine or sides of your bowl of water, you should notice that the waves bounce off and travel back in the opposite direction – this is called *reflection*. You might even notice that the reflected waves can pass through the other waves travelling in the opposite direction. If you remove the jelly babies from one end of the machine, or place an object in the bowl so that the water in one end is much shallower than the other, you might notice that the speed and wavelength of the waves changes.

Having this knowledge of the behaviour of waves on water, which we can directly see for ourselves, allows us to use waves as a *model* for light and sound, because they seem to behave in the same way. So, for example, using the wave model, the idea of frequency allows us to explain why we have sounds of different pitches. All sounds are made by something vibrating, whether it is a guitar string when it is plucked or the skin of a drum when it is hit. If an object is making a sound, some part of it must be vibrating. These vibrations are passed on from the object to the surrounding air, and then spread out as waves through the air, in a way that is similar, but not identical, to the ripples on a pond.

We hear sounds when these waves reach our ears and make our ear drums vibrate, creating signals in our brain. The quicker the guitar string or drum skin vibrates (the higher the frequency of vibration), the higher the note it produces. Similarly, we can model the different colours of light as light waves with different frequencies (or wavelengths). We can also explain reflection and refraction of light, and the fact that it spreads out when it goes through a tiny gap, in exactly the same way we would explain this behaviour for waves on water.

There is a problem with this model that baffled scientists for a long time, and is often spotted by students when these ideas are presented in class: water waves travel through water, sound waves travel through the air (as well as solids and liquids), but light waves from the Sun seem to travel through empty space, so what exactly is vibrating?

Water and sound waves travel through 'stuff', or to use the correct scientific terminology, a *medium*. Before the late 1800s, many scientists thought there *must* be a medium that light waves travel through, because it was impossible for waves to travel through completely empty space. They called this hypothetical substance the *luminiferous ('light-carrying') ether* and many scientists tried, but failed, to prove it existed. The solution to the mystery of how light waves travelled through empty space was found in understanding electric and magnetic fields.

You've probably come across the idea of force fields in science-fiction books and films, but you may be unaware that we spend our entire lives inside a force field – the gravitational field produced by the Earth.

A gravitational field is the term used to describe the region of space around an object with mass in which another object with mass will experience a gravitational force (more about that, and forces in general, in the next chapter). Similarly, a magnetic field is the space around a magnetic object in which another magnetic object will experience a magnetic force, and an electric field is a region of space around an electrically charged object in which another charged object will experience an electric force.

You can't see fields, they are invisible, but you can see and feel their effects. If you have ever played with magnets, you have observed the effects of a magnetic field and how it allows one magnet to attract or repel another magnet from a distance. If you have ever rubbed a balloon to get it to stick to the wall or pick up small bits of paper, you have felt the effects of an electric field. It's important to know about the existence of fields because they help complete the wave model of light.

In 1865, the Scottish mathematician and scientist James Clerk Maxwell showed, using mathematical models, that light waves were 'ripples', which spread out through electric and magnetic fields. Maxwell made one of the greatest breakthroughs in the history of science by helping to prove that electricity, magnetism and light were all interconnected, different aspects of the same thing. According to his work, visible light is made up of electromagnetic waves, which travel through interconnected electric and magnetic fields that spread throughout all of space whenever an electrically charged object, like an electron, is made to move. Just as the different colours of light we can see are different parts of the visible spectrum, visible light itself is part of a bigger *electromagnetic spectrum* that, in order of decreasing wavelengths, includes radio waves, microwaves, infra-red, visible, ultraviolet, X-rays and gamma rays. (I remember this order using the mnemonic Rabbits Mate In Very Unusual X-rated Gardens and my students make up their own, often ruder, ones.)

Modelling visible light as an electromagnetic wave gives us a more sophisticated and useful way of explaining the scattering that leads to a blue sky. Each molecule of gas in the air contains charged particles. When light waves hit them, these charged particles are forced to move, so that they absorb the incoming waves then send out new ones. In effect, the molecules in the air become little vibrating light sources themselves, sending light waves out across the sky where they are absorbed again by other molecules, which then emit their own light waves so that light reaches us from all across the sky, and not just directly from the Sun.

Something that waves do, which things such as footballs and buses do not do, is pass through each other when they cross paths, and carry on going as before. Some interesting things happen at precisely the point where they meet – if a wave crest from one wave meets the crest of another wave on water, the two waves 'add up' at that point and produce a taller crest. Similarly, if the crest of one wave meets the trough of another wave of the same size, they 'cancel' out at that point and the result is that the water there remains still and does not move up or down. This effect is known as *superposition,* and can be used to explain how the colours form on the surface of a soap bubble and also the bright and dark rings of light that can sometimes be seen around a street light in the mist.

The wave model of light is incredibly useful, because it allows us to explain a lot of the *behaviour* of light. However, there are some things light does that cannot be explained using the wave model, but can be explained using the idea that light is made up of 'photons' – tiny particles that carry energy. The two models are connected by the fact that the energy of a photon of a particular colour of light depends on its wavelength –

longer wavelength light has photons which carry less energy, while shorter wavelength light has photons with more energy. This idea was first put forward by Albert Einstein and it led to a revolutionary new model of the world known as *Quantum Mechanics*.

If this was a different book, I would now start telling you about how, according to quantum mechanics, we might be living in one of millions of parallel universes, or that a cat in a box full of poison might be alive and dead at the same time . . . but I hope you don't mind if I don't. Instead, I want to make the point again that we don't *really* know what light is, as it behaves like a wave in some situations and it behaves like a particle in others, which is what is meant by the term 'wave-particle duality'. But while we may not have a perfect idea of what light is, we do have some very good scientific models, which allow us to explain everything from how rainbows form to why the stars sparkle in the night sky.

These models are also what have allowed us to build things such as lasers and flat-screen televisions. I hope that, having read this chapter, you might feel you have a better understanding of why the sky is blue, but I know that you've probably got lots more questions. I've tried to give you an answer to the question 'Why is the sky blue?' that is satisfactory if you started off with little or no knowledge of what scientists think about light. But the nature of science, and indeed the nature of how we respond to answers to these types of questions, is that there is usually more we want to know.

Just as my three-year-old can repeatedly ask 'Why, why, why?', so it is with science – we can always look for another, deeper layer of explanation, a more powerful model that provides even better predictions and answers.

Chapter 2
Why Don't Things Fall Up?

Even if you weren't paying much attention in your school science lessons, I'm willing to bet that you remember that things fall *down* because of gravity. Perhaps you even have some recollection of learning about Sir Isaac Newton, and how a falling apple inspired him to work out the details of exactly how gravity works. But do you remember if your teacher told you that as an apple falls *down* to the Earth, the Earth also moves *up* towards the apple? In which case you might already know that one valid answer to the question 'Why don't things fall up?' is that they do, we just don't notice it.

The story about the apple may not be strictly accurate, but what is undoubtedly true is that Newton was somebody who looked at the world around him and wanted to know why things behaved the way they did. The question at the start of this chapter, 'Why don't things fall up?', is one that I suspect lots of children have asked. It might be sufficient to say 'because gravity pulls things down', but that's probably not entirely satisfying, particularly to older children.

A related question might be: 'Why don't people on the other side of the world fall off?' The answer to this is that the Earth's gravity pulls things towards it, so somebody on the other side of the Earth is pulled towards the ground over there in the same way we are pulled towards the ground here. In this sense, 'down' is merely the name we give to the direction things fall due to gravity. But what exactly is gravity? Newton and other scientists have written entire books addressing this question, and you may be surprised to learn that scientists still don't have a completely satisfactory answer, but I'll do my best to cover some of what they do know in the following pages.

Gravity is a force. But what exactly is a force?

A simple description of a force is that it is a push, a pull, or maybe a twist. And that's not a bad place to start. It can seem as if forces are things that are caused by *people* for the simple reason that, a lot of the time, things around us happen because someone makes them happen. But an apple will fall from a tree without any help from a human, because forces are exerted by *objects,* not just people.

This is the definition of a force I like to use with my own students:

> *A force is an interaction between two objects which can cause one or both of the objects to start or stop moving, change shape, speed up, slow down, change direction or any combination of these things.*

In everyday life, we are most familiar with what are known as *contact* forces, a push, a pull, or a twist that arises when objects are touching. We can also see the effects of *non-contact* forces at work, like gravity, magnetism and the electrostatic force that makes your hair stand on end when you take off a polyester jumper. Non-contact forces act without direct contact between objects involved in the interaction, and are sometimes said to 'act at a distance' – something you can easily see by moving a magnet slowly towards another magnet.

One of the key differences between magnetism and gravity is that magnets can push or pull each other, whereas the force of gravity, as far as we know, is only ever attractive. If I had not learned about it at school (or had not been paying attention when I did), I might have thought that gravity was simply something that Earth, and maybe other planets, had in order to keep things from flying off into space. What is not obvious is that the force of gravity exists between *any* two objects. (More precisely, it exists between any two objects with *mass*, and I'll explain this later.)

There is nothing to indicate to most of us that there is a force of gravity acting between you and this book, or in fact between you and *every single other object in the universe*. But Newton showed that the force that pulls an apple to the Earth is the same force that keeps the Moon in orbit around the Earth, and Earth and other planets in orbit around the Sun. He also showed that if the mass of the objects is small, the force is tiny, and that it quickly gets smaller if the objects move further away, which explains why you feel the pull of gravity from the Earth, but not the pull of gravity from this book or from Jupiter.

Newton summed all this up in what is known as his *Law of Universal Gravitation*, which can be stated as:

> *Every object attracts every other object in the uni-*
> *verse with a force which is directly proportional to*
> *the product of their masses and inversely propor-*
> *tional to the square of the distance between them.*

There's quite a lot of information in that sentence, with words that have specific scientific and mathematical meanings, so let me break it down a little:

- a *law* in science is generally a rule that nature seems to follow, usually based on observations of how things behave. This particular law represents a mathematical relationship between numbers (quantities) for the force, the mass and the distance. Don't worry, I'm not going to ask you to do any maths, but I do think that it's worth seeing how the words above can be expressed as an equation:

$$F = \frac{Gm_1m_2}{r^2}$$

- the *distance* (r in the equation) is perhaps the obvious bit, that is, how far apart the objects are, measured from their centres.

- the *force* (F) is a measure of how much gravitational pull one object has on the other object.
- the *mass* (m) is a bit more complicated, so I'll discuss it in more detail below.
- the *product* of two numbers is simply what you get when you multiply them together.
- two numbers are *proportional* to each other if they increase (or decrease) together in such a way that halving, doubling or tripling one leads to the same happening to the other. For example, if you get paid for every hour of work you do, your wages are *proportional* to the number of hours you work, so you would earn twice as much money if you worked twice as many hours.
- *inversely proportional* means that if one number goes up, then the other goes down in such a way that doubling one quantity halves the other, multiplying one number by four leads to dividing the other number by four, and so on. For instance, the time taken to do a task at work might be inversely proportional to the number of people doing it, so if you double the number of people, the time taken to do the job is halved.
- the *square of the distance* means that the number that represents the distance is to be multiplied by

itself. In this case, it means that moving, for example, three times as far away from an object makes the gravitational pull nine times weaker (three squared is nine).

You might notice there's a capital G in the *equation* that isn't mentioned in the *statement* of the law above – this represents the *universal gravitational constant*, a minuscule quantity with a value of about 0.00000000007 (7×10^{-11}). It is needed to convert the proportional relationship between force, mass and distance into a mathematical equation, and ensures the correct numerical value when you put in the numbers for the masses (in kilograms) and the distance (in metres) to get the force (in newtons). If we measured mass in ounces, and distance in inches, the value of the constant would be different. The tiny value of G is indicative of the fact that only very massive objects have a noticeable gravitational pull, and the reason why we don't feel gravitationally attracted to each other.

Mass

I want to focus for a moment on the word 'mass' in Newton's law, because it is often used interchangeably with 'weight' in everyday life. Most people would get what you meant whether you told them your *weight* was 70 kilograms or your *mass* was 70 kilograms, but strictly speaking (and as a physics teacher I need to be *very* strict on this particular matter), 'mass' and 'weight' have two distinct meanings and should not be used to describe the same thing.

The mass of something can be thought of as the amount of 'stuff' or matter in it. More precisely, the mass of an object is a measure of how difficult it is to *change its motion* – the more mass something has (which scientists call being more massive), the harder it is to start it moving, change its direction when it is moving, or stop it moving. In everyday conversation, we might use the word 'massive' to describe something really big (or maybe your local gang), but scientists can use the word 'massive' to talk about something smaller than an atom, to convey the fact that it has a mass.

Scientists measure mass in kilograms. Your mass would be the same whether you measured it here, or on the Moon or the Sun. Your weight, however, would be about six times less on the Moon and almost thirty times more on the Sun. This is because 'weight' is the word scientists use for the force of gravity on an object and, as Newton's law of gravitation tells us, the size of the force of gravity on you depends on the mass of the object that is pulling you, and how far away from it you are. In science, weight is not measured in kilograms (kg), but in the unit we use for force, the newton (N).

On the surface of the Earth, each kilogram of mass is pulled down by the Earth with a force of about ten newtons, so to get your weight in newtons you multiply your mass in kilograms by ten. In a delightful coincidence, it turns out that an average apple on Earth has a weight of roughly one newton.

I'm a down-to-earth kind of chap, who appreciates the gravity of the situation we have in this country with obesity, so I don't think it would be appropriate to make a light-hearted comment about how we could all achieve weight loss by going to the Moon . . . sorry, I know that this sort of humour is painful, but awful punning is a core requirement of passing most teacher training courses, and if you do understand why going to the Moon would mean you would lose *weight*, then we're making progress.

As I've said, I don't expect you to do any complicated maths while reading this book, but there are some equations that even young children are taught in school and it feels like this is a good place to introduce you to one that I hope is relatively straightforward:

$$W = mg$$

This is the equation for calculating the weight of an object anywhere in the universe. W is the weight of the object, m is its mass and the g is the gravitational field strength, which is the size of the force pulling down on each kilogram of an object at that point. I mentioned in the previous chapter that a gravitational field is the region of space around an object with mass in which another object with mass will experience a force. Technically, even though it gets weaker with distance, the gravitational field around an object goes on forever. This is why even though the Moon is nearly 400,000 kilometres away, its gravity can still pull with enough force to cause the tides on Earth.

I'm not going to go into detail about how Newton came up with his law of gravitation – there are plenty of other books which explain this – but there is one thing that I think is worth knowing about what he did. First, although Newton was undoubtedly smarter than you and me in some respects – he had the kind of brain that could see patterns and do mathematics in ways you or I would probably struggle with – he would not have been able to do what he did without the work of other scientists. This is important because, while scientific discoveries are often credited to individuals, science is really a group activity, something that requires collaboration and the sharing of knowledge.

In the case of Newton's universal law of gravitation, Newton was able to draw on the work of Johannes Kepler, who had come up with three laws of planetary motion, which showed there were mathematical relationships that described the orbits of the planets around the Sun. Kepler himself had used observations collected by his boss, Tycho Brahe, to formulate these laws, and Brahe will have used the work of others to do his work, and so on. Newton himself was a complicated character and in some ways unpleasant as a human being. But to his credit he acknowledged the role of other scientists in his own work, saying, 'If I have seen further, it is by standing on the shoulders of giants.'

I have heard scientists say that Newton's law of universal gravitation should rank alongside the complete works of Shakespeare or the building of the pyramids or Michelangelo's painting of the Sistine Chapel as one of humanity's greatest cultural achievements.

But it was not just gravity that Newton explained, it was how forces in general work, which is why we measure force in a unit named after him.

Laws of Motion

Something you are unlikely to have learned at school, unless you did A-level Physics, is that there are actually only four known forces in nature – gravity, electromagnetism, the strong nuclear force and the weak nuclear force. Nearly *everything* we see happening in the world around us is down to only two of these – electromagnetism and gravity. Most people are somewhat familiar with the effects of gravity, but relatively few know that every push, pull and twist, every occurrence of friction, air resistance, magnetism, electricity and every single chemical reaction that takes place is, at its root, caused by electromagnetism.

As their names imply, the strong nuclear force and the weak nuclear force are responsible only for things that happen within the nucleus of an atom. The strong nuclear force is needed to hold protons and neutrons together in the nucleus, while the weak nuclear force is often responsible for subatomic particles breaking apart. (I'll come back to atoms and what they're made of in a later chapter.)

Now, let's get back to what you should have been told at school and some basic rules about how forces behave, known as *Newton's Laws of Motion*. These laws were first published in a book by Isaac Newton in 1687, and are now widely misunderstood and misremembered by pretty much everyone who encounters them.

I'll start with Newton's Third Law, which people often quote as 'every action has an equal and opposite reaction'. I'm not a big fan of this way of expressing it and prefer to state it like this:

If an object, A, exerts a force on another object, B, then object B exerts an equal-sized force in the opposite direction on object A.

Admittedly, this doesn't have quite the same ring to it as the commonly known statement, but I think learning this version makes it much easier to apply Newton's Third Law *correctly* when attempting to understand how the world works. The law tells us that *forces always occur in pairs*. But this *does not* mean that one force causes the other one to come into existence, or that forces always cancel each other out, or that they have the same effect on both objects, as implied by the statement 'every action has an equal and opposite reaction'.

Here's a really simple way to see Newton's Third Law in action for yourself:

Find a friend who is much heavier or lighter than you, and a couple of skateboards (or wheelie chairs if you've relinquished your boards and given up trying to hold on to your youth). Stand on a skateboard and tell your friend to do the same. Now try pushing your friend away while you stay still. You should find that this is impossible to do. When you push your friend away, you will also move away from your starting position, in the opposite direction.

You should now be able to explain why: because if an object, A (you), exerts a force on another object, B (your friend), then object B (your friend) exerts an equal-sized force in the opposite direction on object A (you). You should also notice that if your friend is heavier than you, they move away more slowly, and if they are lighter than you, they move away more quickly. The reason for this is explained by Newton's Second Law, which I'll come on to soon.

Newton's Third Law can be seen in action whenever something is moving – we walk by pushing backwards on the ground, birds fly up by pushing air down, and anything that swims can only move forward by pushing water backwards (this is particularly easy to see if you do the breast stroke). Once you understand this law, you can use it to explain everything from how rockets work or why a gun recoils when you fire it, to why you need to push *down* on the ground to jump *up*. Newton's Third Law also explains why every apple being pulled down to the Earth is also pulling the Earth upwards.

The main reason we do not notice the force of an apple acting on the Earth is due to *Newton's Second Law*, which you might have learned in school as the equation $F = ma$ (force equals mass times acceleration). I've found that, for some reason, this is one of the few equations that people recall from their school science lessons, but many would be pushed to explain that it means:

> *An unbalanced force will make an object accelerate and the acceleration will be proportional to, and in the same direction as, the unbalanced force.*

The word 'unbalanced' is important here, because objects often have more than one force acting on them, and in more than one direction. What I have to teach my students is that the 'F' in $F = ma$ is the 'resultant' force on an object. This can be thought of as meaning the 'overall' force on an object, but resultant has a more precise mathematical meaning, which I'll try to explain in a little more detail.

If you have ever tried to push-start a car, you might have been unable to do it until someone else helped you. It might be obvious that you needed more force to make the car move, but what is perhaps not obvious is *why* you need more force. It's not just because the car is heavy (or, more correctly, 'massive'). The reason why you might not be able to get a car moving by pushing it on your own is that there is another force acting in the *opposite* direction to your push: *friction*.

Friction is a really interesting force – without it we would not be able to walk, it's the reason why your hands warm up when you rub them and it's why the tyres on a car and countless other things wear out as we use them. But for now, all you need to know is that friction always acts in the opposite *direction* to motion, so if you're trying to push a car *forwards*, friction is effectively pushing *backwards*.

I hope you can see that if you are pushing the car by yourself and the car is not moving, your push is cancelled out or *balanced* by the opposing force of friction. To get the car to move, the force pushing the car forwards must be greater than the maximum frictional force, in other words the forces must be *un*balanced. When this is the case, we say that there is a *resultant* force on the car. The resultant force on an object is the overall force on an object when the sizes and directions of all the forces on it have been taken into consideration.

Newton's Second Law tells us that if the forces on an object are unbalanced (the resultant force on it is not zero), it will *accelerate*. We usually think of 'accelerating' as an increase in speed, but in the context of Newton's Second Law (and in science in general), the word *acceleration* is used to describe any change in the motion of an object – speeding up, slowing down, or changing direction. So, if you're walking at a steady speed but going round in circles (constantly changing direction) then you are accelerating.

Newton's Second Law ($F = ma$, i.e., the resultant force on an object is equal to its mass multiplied by its acceleration) tells us that the acceleration, or change in motion of an object, is proportional to the size of the resultant force acting on it – the bigger this force, the bigger the acceleration. This explains the difference in behaviour between a ball you've thrown gently and one you have thrown with all your might.

This law also tells us that *for a force of a particular size,* the acceleration is *inversely* proportional to the mass of an object – the bigger the mass, the smaller the acceleration. (You might remember that you can *rearrange* the equation F = ma to get $a=\frac{F}{m}$, i.e., acceleration equals the resultant force divided by mass, which shows this mathematically.) This is why throwing a football and throwing a cannon-ball with the same force would not produce the same results, and why a car accelerates more slowly when it is full of people. Coming back to that apple of Newton's, his Second Law tells us that even though the same *sized* force is acting on the Earth as on the apple, the resulting movement of the Earth (which we can calculate using $F = ma$) is less than the width of an atom.

Let's move on to Newton's First Law now. Before I tell you about it, I'd like you to answer a question that anyone who has studied school physics *ought* to get right:

Imagine a football that has been kicked in the air and is approaching the highest point it will reach before coming back down. Now, ignoring air resistance, which of the following statements correctly describes the forces acting on the ball at this point?

A: There is only one force acting on the ball, pushing it in the direction it's moving.

B: There is only one force acting on the ball, pulling it downwards.

C: There are two forces acting on the ball, one pushing it forwards, and one pulling it downwards.

The correct answer is B. If you got this right, and not by luck, I suspect it's because you know and, more importantly, understand Newton's First Law, which can be expressed as:

> *A stationary object will stay at rest, and a moving object will continue moving in a straight line, at the same steady speed, unless the forces acting on them become unbalanced.*

I have put the above question to entire lecture theatres full of people with A-levels and even degrees in science and they have got it wrong. Most people pick C. The reason for this is a common misconception – that an object always needs a force acting on it to propel it forward. However, Newton's First Law tells us that there is no need for such a force; a ball that has been kicked should carry on moving in a straight line at a steady speed forever.

The reason why we find it so hard to believe this, even if we have been taught it properly at school, is that it's something we cannot easily see for ourselves here on Earth and it seems to be strongly contradicted by our everyday experiences – to keep a car moving we need to keep our foot on the accelerator, to keep a bike moving we need to keep pedalling, to keep a skateboard moving we need to keep pushing. The reason for this is that there are always forces, like friction and air resistance, which slow objects down or cause them to change direction. But you can perhaps now imagine that if there was not any gravity, or air resistance, and you threw a ball into the air, it would carry on moving at a steady speed, in a straight line forever.

Newton's First Law is not as easy to demonstrate as his Third Law because it's usually quite difficult to completely get rid of friction and air resistance in real life situations. In their attempts to convince their students that this law is actually true, many physics teachers around the world have battled with a contraption known as an 'air track'. A vacuum cleaner is used in reverse to blow through tiny holes in a metal track, with the aim of creating a cushion of air on which small trolleys can float and travel without experiencing much friction. If you didn't see this at school, you will have seen a similar thing in action if you've ever played air hockey. When they work properly (they seem to be easy to bend out of shape and need to be perfectly level), air tracks can indeed show something moving in a straight line at a steady speed. However, it's only for a very short distance, and I'm not sure it's entirely convincing for many students. It's always worth trying though, because there are many situations in life that make sense once you get your head round this particular law of physics. For example, the reason why you feel thrown forward when you're in a moving bus or train that suddenly comes to a stop is not because there's suddenly a force pushing you forward, but because you *carry on moving* when the vehicle comes to a stop. Similarly, the reason why you feel flung to the side of a car when it turns a corner is that your body continues moving in a straight line until there is a force acting on it to change its direction and make it go in the same new direction as the car.

Finally, here's an example of Newton's First Law in action which answers a tricky question I've heard asked by many children, and indeed adults: if the Earth is spinning, why doesn't it spin away from us when we jump into the air? The reason is that when we jump *up* in the air, we still continue moving *horizontally* at the same speed as the Earth below us because there is no resultant force to change our speed in that direction.

To really understand and be able to apply Newton's First Law, we need to do something that is often necessary to arrive at a *scientific* understanding of the world: overcome our intuitive, common sense ideas of how the world works. It's natural to think that if something is moving, it *must* have a force propelling it, and to believe that if something is not moving, it *doesn't* have any forces acting on it.

Newton's laws tell us both these ideas are incorrect. Newton didn't have air tracks or computer simulations to show him how the world behaved without friction or air-resistance, instead he was able to *imagine* what it would be like without these things and see what was really going on. This doesn't come naturally to most of us, and is why it often takes a great deal of effort to be able to apply his ideas ourselves when trying to explain natural phenomena.

In my experience as a teacher, true understanding of school science comes with not just being able to regurgitate laws and equations on demand, but through *practising* applying them to unfamiliar situations. I have taught very few students who could immediately take in what was presented in class and then be guaranteed to answer correctly any exam question they might face. Even the brightest of students usually need to spend some time doing question after question that requires them to *use* what they've learned in class. Students often complain about this, but testing our understanding of scientific models by trying to explain how things behave in different situations is really the only way to be good at science.

Falling

So, here's a question for you to try: if you had a heavy bowling ball and a much lighter football of the same size, and dropped them both from the same height, which one would land first? Why?

I once filmed this experiment for a television programme, with the presenter dropping the two balls from a tall tower. Once we had finished filming, the cameraman said: 'We'll have to do that again, both balls landed at the same time!' He was mystified when I told him that no, we had got exactly the shot we needed, that I wanted to show that a heavy ball and a much lighter ball will fall at the same speed. My cameraman was not being stupid, he was simply responding to his gut instinct, which tells most of us that a heavy object will reach the ground more quickly than a less heavy one.

You may agree with this: after all, surely a feather will fall much more slowly than a stone? It's something you can investigate for yourself:

> *Take a tennis ball or something similar (a pen will do) and a flat piece of A4 paper and drop them from the same height, of at least a metre or so. You should find that the ball lands first while the piece of paper takes its time, gently wafting down.*
>
> *Now, try the same thing again, but this time, screw the piece of paper up tightly into a ball first. You should notice that the ball (or whatever you're using) and paper land pretty much at the same time. The difference is that the screwed-up paper is much less affected by air resistance, the force due to air particles hitting it, because it has a smaller surface area than when it is flat.*

If you could do this experiment in a room without any air in it, you would find that the paper and ball fall at the same time, even when the paper is flat. This experiment was famously done with a feather and hammer by astronauts on the Moon and if you've got the time, it's well worth looking up the video footage of it on the internet.

So, why do objects dropped from the same height land at the same time if there is no air resistance? The answer lies in a couple of the ideas I've already discussed. Remember that the force pulling objects down is gravity, and that the size of the gravitational force on an object is called its 'weight'. The more mass an object has, the greater its weight, so the more strongly it is pulled towards the Earth.

This might make you think that heavier objects ought to be pulled down to the Earth faster. But remember also that the acceleration of an object, according to Newton's Second Law, is *inversely* proportional to its mass – so an object with a bigger mass needs a bigger force to give it the same acceleration as an object with smaller mass. In effect, being *heavier*, having a greater weight, is cancelled out by being more *massive*, so a heavy object and a light object end up falling at the same rate.

In fact, if you've been paying attention, you will want to correct me and point out that what I meant to write is that:

All objects, regardless of their mass, experience the same acceleration due to gravity.

The Earth's gravity means that all objects near the Earth's surface accelerate at about 10 metres per second squared – what this means is that any object dropped from a height will speed up by 10 metres per second every second. So after one second it will be travelling at 10 metres per second, after two seconds its speed will be 20 metres per second, after three seconds it will be 30 metres per second, and so on until it hits the ground.

It's not easy to see that falling objects get faster as they fall in everyday life, but if you have a slow-motion feature on your camera phone, try this: film any falling object from a height of at least 2 metres (the higher the better) and then look at the resulting slow-motion film. You should just be able to tell that the object falls faster towards the end of the film. If your camera allows it, scroll through the footage and see how far it falls in every frame – you should see that it falls a greater distance in every successive frame.

This acceleration due to gravity is given the symbol little 'g'. Ah, but isn't that the symbol for gravitational field strength? Why yes, it is. Can you see why they have the same symbol and same numerical value? It's one of those things that is very satisfying for students to see when I do it on the board at school, so let me do it for you here:

Remember, the force due to gravity (an object's weight) is given by the equation $W = mg$.
And that the general equation for a force on an object is given by $F = ma$.

I hope you can see that these equations are equivalent to each other, that W and F are both a force, so that means mg and ma are the same thing. Therefore, we can write $mg = ma$. Since m (the mass) is the same on both sides, g and a must also be the same or the equation wouldn't be true. In other words, the number we calculate for gravitational field strength using Newton's laws, 10 newtons per kilogram, is the same number we get for the acceleration due to gravity, 10 metres per second per second.

Most of us (fortunately) don't ever fall from high enough to experience the acceleration due to gravity for very long. It's worth comparing it to the maximum acceleration of a modern family car, which is about 5 metres per second squared and to note that even a

Formula One car can speed up at a maximum of about 15 metres per second squared. Considering this, you may believe, as some people do, that if a penny was dropped from a tall building and it hit you on the head, it would kill you. The tallest building in the world at the time I am writing this is the 828-metre tall Burj Khalifa in Dubai, although it is due to be surpassed in height by the Jeddah Tower in Saudi Arabia, which is planned to be the world's first kilometre (1,000 metre) tall skyscraper.

Now, I've done the calculations, so that you don't have to: A penny dropped from a height of 1,000 metres, getting faster and faster by 10 metres per second, every second, due to the acceleration caused by gravity, would take more than 14 seconds to hit the ground and be travelling at about 140 metres per second (313 miles per hour) when it did. Depending on whether the penny landed flat or on its edge, it could well do some very serious damage to your head and kill you.

However, a dropped penny would not make for a good murder weapon, because no penny dropped from even a much higher building would ever reach a speed of 140 metres per second. As you've probably guessed, the reason for this is air resistance, which acts in the opposite direction to gravity. Air resistance also increases as an object moves faster so that, for a falling object, at some point the size of the force due to air resistance is exactly the same as the force due to gravity, so they become balanced. As Newton's Second Law tells us, a balanced force does not cause any acceleration, so the falling object stops getting faster and falls at a constant speed.

This maximum steady speed is called the *terminal velocity*, which is a term you might have come across (I'm sure there is at least one action movie with 'terminal velocity' in the title). The terminal velocity a falling object reaches depends on it shape and size, so while a flat penny dropped from the top of a building might not kill you, a long, thin object might.

Okay, so we've established that forces, including gravity, make things accelerate, but only if they are *unbalanced*. If you can remember a couple of pages back, I explained that 'accelerate' does not always mean making things go faster, it can also mean slowing things down or making them change direction. This is important, because it should help you answer the final question I want to deal with in this chapter – if gravity makes things fall down, why doesn't the Moon fall onto the Earth?

To answer this question, it helps to do a thought experiment similar to one that Newton carried out himself:

Imagine you're standing on top of a mountain, and you've got a rifle and some bullets. You press the rifle against your shoulder and fire a bullet away from you, so it leaves you in a perfectly horizontal direction. Once the bullet has left the gun, there is no force propelling it forwards, and for the purposes of this thought experiment, we'll imagine that there is no air resistance slowing the bullet down. However, the bullet will fall downwards due to gravity, and that will mean its path will be curved.

Now, imagine you have a more powerful rifle, one that can fire the bullet out at a much greater speed. This time the bullet will still have a curved path, but it will have travelled much further away from you by the time it hits the ground. I hope you can see where I'm going with this . . .

Imagine you had a rifle that could fire a bullet so fast that the path of the bullet curved so that it exactly matched the curvature of the Earth – the bullet would never hit the ground but instead keep going round and round the Earth. In other words, the bullet would be in orbit around the Earth. In this special case of the bullet fired at exactly the right speed to make this happen, the force of gravity changes the direction of the bullet but not its

speed (how many metres it travels every second),
so that it continues travelling at the same speed
while constantly 'falling' towards the Earth.

This is pretty much how the satellites we use for communications, GPS, weather forecasting, and so on are put into orbit. They are taken high above the Earth's surface and 'fired' with just the right speed to make them go into an orbit. In science, 'satellite' means any object that orbits another one, so the Earth and other planets are satellites of the Sun, and the Moon is a satellite of the Earth. So the answer to the question, 'Why doesn't the Moon fall onto the Earth?' is that the Moon *is* constantly falling towards the Earth, but it is moving through space at precisely the right speed to ensure that the surface of the Earth always curves away from it just enough, so that it never hits it.

I hope this chapter has jogged some memories and perhaps clarified some ideas about forces for you. Even if we can't all do calculations or answer questions about the movement of satellites or planets, isn't it wonderful to know that today, long after Newton developed these ideas, people can use them to send probes out deep into space so that we can see further into the universe than he could ever have hoped or imagined?

Chapter 3
Why Does Ice Cream Melt?

There are many joyous moments watching a baby grow up, particularly all the 'firsts' – the first time they smile properly, the first time they say an actual word, the first time they projectile poo over you, and so on. A friend told me that I should make sure to watch my daughter's face the first time she ate ice cream, and I wasn't disappointed: there was an initial look of surprise from the cold before a massive smile burst across her face as the ice cream melted in her mouth and the sweet, creamy mango flavour hit her. She was so delighted that she literally clapped her hands with glee, as if she understood that this miraculous substance was one of life's greatest gifts and deserved a round of applause.

A less fun thing to do as a parent is to watch a child eating ice cream on a hot day, while it melts and drips all over them. If I'm treating my children to an ice cream while out, I try to avoid the 'soft-serve' stuff, because it melts so quickly and virtually guarantees needing a change of clothes when we get home. But why does ice cream, or an ice lolly, or chocolate, or *anything* melt? The answer to this lies in what some scientists have said is the most important idea in science – *the particle model of matter*.

The word 'particle' is often used to describe a small, *separate*, bit of something, like a grain or a speck. For the particle model of matter, it's useful to imagine particles as tiny balls that are the absolute smallest bits of a substance. You probably know that these particles are called *atoms* and *molecules*, and that they are not necessarily spherical, but for now we don't need to worry about this. Remember, this is a *model*, a simplified version of reality that we can use to explain and predict things in the real world. To get to grips with this particular model, and be able to *use it*, you need to remember the following:

- all matter, all the stuff around you, is made up of tiny, tiny particles which are too small to be seen individually – one grain of sugar contains about 1,000,000,000,000,000,000 particles of sugar, which means there are millions and millions of times more particles in a single grain of sugar than there are grains of sugar in a 1 kilogram bag of it

- these particles are *always* moving (which is why the particle model is also called the *kinetic* theory of matter, because 'kinetic' refers to moving things)
- the hotter a substance is, the faster its particles are moving
- the particles attract each other
- there is always some *empty space* between the particles

Temperature is a key part of the particle model. In everyday life, we use the idea of temperature to tell us whether things are hot or cold – the higher the temperature of something, the hotter it is. In the particle model, the temperature of a substance is directly related to how fast its particles are moving – the higher the temperature, the higher the *average speed* of the particles. Weather reports, recipes and other everyday measurements of temperature use the Celsius scale (also known as the centigrade scale), where 0°C is the freezing point of water and 100°C is its boiling point. Very cold temperatures have a negative value on this scale, for example the temperature in a standard kitchen freezer is around minus 18°C.

In science, temperature is measured using the Kelvin scale, which does not have negative values. Zero kelvin (0 K) is known as 'absolute zero', because it is the lowest possible temperature anything can have. It is equivalent to minus 273.15°C, and is the temperature at which, according to the theory, particles completely stop moving.

It might not be obvious, but by using these ideas, we can explain a lot of things that we see in nature, including why ice cream melts, and why it melts more quickly on a hot day than a cold day. We can also explain why you can smell someone's perfume from across a room, why condensation forms on a cold surface, why warming up syrup makes it easier to pour and why you can make the lid on a jar easier to open by running it under hot water first.

Most substances that we encounter regularly on Earth are found in only one state of matter, existing either as a solid, liquid or gas. Water is unusual in that it can be found in all three of these states in your kitchen – solid in your freezer, liquid in your taps and gas coming out of your boiling kettle. You can't actually *see* water when it's a gas because, like most gases, it's invisible. The cloud you see when a kettle boils is made up of tiny droplets of *liquid* water. But next time you see a kettle boiling, look closely at it and you should notice that there is an apparently empty gap just above the spout and below the cloud – it is in this gap where the water is a gas.

Water is the obvious substance to use when talking about solids, liquids and gases, but I also like to use a burning candle to show my students the three states of matter. They can see that the candle itself is made of solid wax, and that there is a pool of liquid wax that forms below the wick. I take a lit match and hold it against the solid wax, and then against the liquid wax, to show that neither catch fire – we can all be sure that wax does *not* burn when it is a solid or a liquid.

So, what is burning? Most students will respond that it's the wick. But if it's only the wick that is burning, why do you need the wax? One answer I've often heard from students is that the wax slows down the burning of the wick. This is correct, in a way, but how exactly does the wax do this? The answer to this mystery can be revealed with a lovely demonstration that you can do for yourself at home:

Light a candle and let the flame burn for at least a minute. Then, light a match and have it ready to light the candle again. Blow out the candle

*with a short sharp breath and wait a moment
until you see white smoke rise from the wick.
Place your lit match into this white smoke and
you should see the flame from the match 'jump'
through the air and re-light the candle. This is
quite surprising and impressive if you haven't
seen it before. If you practise it a few times, and
do it in a room where the air is still, you can get
the flame to jump several centimetres.*

The reason why the flame travels from the match to
the wick in this rather spectacular way is that the smoke
above the candle contains gaseous wax, which catches
fire. The flame of a candle is, almost entirely, the result
of this burning gas. The wick of a candle is necessary
because setting fire to it starts the process of changing
the wax from a solid to liquid. The wick also provides
a 'pipe' through which the liquid wax can then travel
into the flame and be turned into the gas.

Changing State

So, what's going on when the wax changes state, from
solid to liquid to gas? It might seem possible that the
particles of wax themselves are changing, but this is
not actually the case:

The particles in a substance stay the same when it changes state, it's simply how they're arranged and how they move that changes.

When something is a solid, the particles in it are packed very closely, and held together very strongly with only small spaces between them. The particles are effectively locked in place, in fixed positions, by the forces of attraction between them. The strength of this attraction between particles varies from substance to substance, which is why you can easily break a thin candle, but not a metal stick of the same size. The particles in a solid are not totally static; they can't move from one place within it to another, but they do constantly vibrate.

Heating up a solid makes the particles in it vibrate faster, and so its temperature increases. To begin with, this will usually make the solid expand. This is not because the particles get bigger, but because their increased vibration means they take up more space. This is more noticeable in some substances than others, but it's the reason why you can loosen up a tight-fitting lid on a jar by running it under some very hot water – the lid gets slightly bigger so it doesn't fit the jar so snugly (the glass jar also gets slightly bigger, but not as much as the lid). This is also why railway tracks need to be laid with gaps between them, so that when they expand in hot weather, they don't buckle.

As you carry on heating a solid, the particles in it vibrate faster and faster until they break away from their fixed positions. The particles become slightly further apart from each other, but each remains in close contact with the ones surrounding it. The forces holding the particles together become slightly weaker and the spaces between them become slightly bigger, so they can now slide past each other and move around from place to place. This process is called *melting* and results in the solid turning into a liquid. In most cases, these changes in the movement and arrangement of particles mean that a liquid takes up slightly more space than a solid, so has a slightly bigger volume than when it was a solid.

However, this does not mean you can squash a liquid to make it smaller – a liquid is effectively *incompressible*, because its particles are still mostly in contact with each other. This property of liquids allows them to be used in hydraulic technologies, used in things such as fork lifts and car braking systems, which allow forces (pushes and pulls) to be transmitted from one place to another through liquids in pipes.

The changes in the movement and arrangement of particles when something melts mean the liquid does not have a fixed shape and can *flow*, taking the shape of whatever container it's poured into. Heating up a liquid makes its particles move around more quickly and further weakens the forces between them, which is why, for example, hot syrup is more runny than cold syrup. If you didn't know it before, you might now be able to see how a mercury or other liquid thermometer works: the volume of the liquid in it changes with temperature, so as it gets warmer it takes up more space in the narrow tube and gives a higher reading, and vice versa.

There's a lovely, simple way to see that the particles in a liquid are in constant motion, and that they move faster when the liquid is hotter:

Take a glass of hot water and a glass of cold water and carefully put a single drop of food colouring into each. You should notice that the food colouring spreads through the water even without the water being stirred, and that it spreads more quickly in the hot water than in the cold water.

If you continue to heat a liquid, its particles eventually shake themselves free so they can completely separate from each other. This process is called *evaporation* and results in the substance becoming a gas. The forces between the particles in a gas are not strong enough to hold them together, so they can move about freely in all directions and will spread out and fill any container they're put in, no matter what shape or how big it is.

The particles now have much, much larger spaces between them than when they were in a liquid or solid – if you could take all the air in a typical school classroom and cool it down into a liquid, it would fit into just a few small buckets. The speed at which particles in a gas move depends on the temperature – the hotter it is, the faster they move. At normal room temperature, about 20°C, the molecules in air are moving at an average speed of around 500 metres per second or so.

Heating up a liquid to its boiling point is the quickest way to make it evaporate and turn into a gas. But you've probably noticed that puddles will dry up even when it's cold outside. The reason for this is that the particles in a liquid are not all moving at the same speed – there are *always* some particles near the surface of any liquid that are moving quickly enough to break free from their neighbours. The hotter the liquid, the more of these fast particles there are, and the quicker it evaporates. This is why, on a warm day, puddles do not last as long and wet clothes dry more quickly – the liquid water becomes water vapour and escapes into the air.

This also explains why you can smell someone's perfume from across a room – the liquid perfume slowly evaporates from their skin, turning into a gas that spreads out into the space around them. We get condensation on cold surfaces because there is always some water vapour in the air, and when these particles come into contact with something cool, such as an ice-cold glass, they slow down and get closer together, forming liquid water droplets.

The particle model explains lots of the properties of gases, like why, unlike liquids, they are *compressible* – if you put your finger over the end of a bicycle pump or syringe you can squash the air in it, because there are relatively big, empty spaces between the particles. If you do this, you can feel that the *pressure* inside the bicycle pump or syringe increases.

The particle model tells us that this pressure exists because the particles of a gas, which are constantly moving about, are continually hitting the surfaces of any container they are in. The faster they move, the harder, and more often, they hit the surfaces. So, if you have a gas trapped in something, like a tin with its lid on, and you continue to heat it up, you can end up with an explosion as the particles move around with more and more energy and hit the walls of the container with greater and greater force and more and more often.

I hope you can see why the particle model is so special – it is tremendously useful for explaining and predicting the behaviour of all sorts of things. So it may surprise you to know that, despite this, not all scientists believed that particles were *real* until the early twentieth century. The problem, according to these scientists, was that they couldn't *see* these particles, so couldn't be *sure* they existed. These days, scientists can use various techniques to produce images of atoms and molecules, and even move individual atoms around to write words or make pictures. But long before we could do this, Albert Einstein helped to convince doubting scientists that the particles we imagine when using the particle theory of matter are real, even if we can't see them.

Einstein is most famous for his equation $E = mc^2$, but in the same year that he published this equation, 1905, he also published a mathematical explanation of a phenomenon that had been puzzling scientists for close to a century: *Brownian motion*.

Proof of Atoms

There is a joke about this that I find absolutely hilarious, but which usually leaves my students staring at me in pity: I hold up a chocolate brownie, jiggle it around and ask, 'What's this?' The answer, which I'm happy to report many students guess, is 'brownie-in-motion'. It's funny because a) it's a pun and puns are always funny, and b) if you move the brownie in just the right way, you can indeed demonstrate Brownian motion. (Okay, maybe it's not *that* funny.)

Actual Brownian motion is usually shown to school students using a microscope to look at smoke particles suspended in air or plastic 'microbeads' suspended in water. In both cases, what you see is tiny specks of stuff randomly jiggling about – this is Brownian motion, named after Robert Brown, who is credited as being the first person to describe this sort of 'jittery' movement of microscopic particles.

In 1827, Brown had been looking at tiny particles, which came out of grains of pollen, suspended in water. These particles did not stay still in water, but continuously jiggled about, even though the water seemed to be absolutely still. He became fascinated by this motion and did further experiments with various other substances, such as glass and metals, which came from non-living sources. Brown found that, as long as he could grind the substance up into a fine enough dust, the particles behaved in the same way when suspended in water.

With these experiments, Brown proved that this constant jiggling about could not be because the particles were alive, as some people had suggested. In his findings, he wrote that 'extremely minute particles of solid matter, whether obtained from organic or inorganic substances, when suspended in pure water, or in some other aqueous fluids, exhibit motions for which I am unable to account'. He was not alone – nobody else at the time could explain this mysterious movement.

Nearly fifty years after Robert Brown died, Albert Einstein published his suggestion for how Brownian motion could arise. The scientific paper, when translated into English, was titled, 'On the movement of small particles suspended in a stationary liquid demanded by the molecular-kinetic theory of heat'.

In the introduction, Einstein wrote that 'it will be shown that according to the molecular-kinetic theory of heat, bodies of microscopically-visible size in a liquid will perform movements of such magnitude that they can be easily observed in a microscope . . . It is possible that the movements to be discussed here are identical with the so-called Brownian molecular motion.'

Even if you could follow and understand the maths in Einstein's paper (I suspect I would struggle with it myself), you might be left wondering exactly why he is often credited as *proving* the existence of atoms. The fact is, Einstein did not provide *direct* proof that atoms or molecules existed. Instead, he did some calculations using established laws of physics, and the *idea* of tiny, invisible particles moving around and hitting a bigger particle, to *predict* Brownian motion *should* occur.

One way to picture what Einstein was suggesting is to imagine what would happen to a giant inflatable ball placed on top of a crowd of people waving their arms around at a pop concert – the ball would jiggle about as people randomly hit it, sometimes making it move a bit this way and sometimes a bit that way, depending on the number of hands hitting it at any instant on any particular side. It would gradually and erratically move over the crowd. The ball represents a big particle and the people's hands are the smaller particles of liquid, or air, randomly hitting it.

Einstein had used mathematics to predict how fast and how far particles in Brownian motion should move. These calculations were put to the test by a French physicist, Jean Baptiste Perrin, who carried out experiments and *measured* the movement of particles. Perrin's work showed that Einstein's calculations were correct and was important in changing the minds of scientists who had been sceptical about the actual existence of particles. The work by Einstein and Perrin effectively settled the argument amongst scientists about whether atoms were real. The fact that Einstein had used the particle model of matter to *predict correctly* things which could be seen and measured meant he was the one who was credited, by some, as proving the existence of atoms.

Both Einstein and Perrin went on to win Nobel Prizes in Physics for their contributions, but Einstein is the name that most people will recognise. I imagine very few people have heard of Perrin, and I suspect even fewer have heard of Marian Smoluchowski, another scientist whose work helped scientists to understand what caused Brownian motion. The reason I mention this is that school science often presents scientific findings as the work of lone geniuses, brilliant people who somehow reveal the workings of nature all by themselves. But this is not how science works – the reality is that scientists work together, learning from and developing each other's work.

Today, schools teach children the particle theory of matter from a young age, and it seems so straightforward that it's easy to forget that this model did not always exist and had to be invented by some of the best scientists who have ever lived. I hope you feel that it provides a satisfactory answer to the question of why ice cream melts – when it's taken out of the freezer, the particles are packed closely together and it's hard to separate them even with the edge of a metal spoon. Then it starts to warm up and the particles vibrate more and more until they shake themselves loose from their fixed positions, eventually ending up smeared over your child's hands, face and clothes.

However, the particle model of matter I've described above can't explain, for example, why ice cream tastes sweet or why it can't be used as a fuel in the same way as candle wax. This simple model also can't explain why some substances, like paper, do not melt when you heat them up but burn instead, or why different substances react in different ways when brought into contact with each other. To be able to explain these things, we need to know a little bit of the science called *chemistry*, and what it tells us about the behaviour of those particles known as *atoms*.

Atoms

The person usually credited with first coming up with the idea of atoms was the Ancient Greek philosopher Democritus. He imagined that any substance could be cut up into smaller and smaller bits, but only up to a point, after which it could not be cut any further. According to Democritus, these last, smallest bits of stuff must be indestructible, and so he called them 'atoms', from the Greek word for 'uncuttable'.

Democritus believed that there were an infinite number of different types of atom, with infinite different shapes and sizes, which made up the seemingly infinite variety of substances in the world. He reasoned that the different behaviour of different substances could be explained by the properties of the atoms from which they were made – atoms of air would be very light, atoms of water would be smooth and slippery, atoms of salt would be sharp and pointy (explaining its taste), and so on. It took a couple of thousand years to get from Democritus's suggestions to the more sophisticated and useful models we now have.

These days, when we talk about 'atoms', we know that they are not indivisible like the ones Democritus imagined, but are made up of three, even smaller, subatomic particles – electrons, protons and neutrons. We also know there are just under a hundred or so different types of naturally occurring atoms that make up everything around us, and Democritus was kind of correct in thinking that the precise properties of these atoms play a role in determining the behaviour of the substances they make up.

One reason why it took so long to arrive at the modern understanding of atoms is that Democritus was a philosopher, someone who tried to come up with answers to big questions such as 'What is everything made of?' simply by *thinking*. He didn't do any experiments to *test* his ideas or provide any *evidence* that his ideas were correct. In other words, Democritus was not a scientist.

By the 1600s though, there were lots of people around the world who were carrying out experiments to try and learn more about how matter behaved and, whilst Democritus can be given credit for coming up with the name 'atom', there is a long list of scientists, including Antoine Lavoisier, John Dalton, Dmitri Mendeleev, Marie Curie, J.J. Thomson and Ernest Rutherford, whose work shaped how we think of atoms today.

The experiments carried out by many of these pioneers of science were similar to the sort of thing you may have done in your chemistry lessons – putting different substances together and seeing what happens when heating, filtering and distilling them or measuring their mass before and after taking part in a chemical reaction. One of the things we are meant to learn early on in our chemistry lessons is that:

Chemical reactions occur when two or more substances are brought into contact with each other and interact to form one or more new substances.

Studying chemical reactions allowed scientists to deduce the properties of atoms in a way that simply thinking about them would not. If you remember putting a piece of metal into an acid and collecting the hydrogen gas that is given off, or making bright blue copper sulphate crystals from black, powdered copper oxide and sulphuric acid, you can perhaps imagine how interesting and exciting it was for scientists doing this without instructions from a textbook, or a worksheet telling them what to expect.

It's often difficult for modern historians of science to establish exactly who was responsible for particular discoveries in the early days of chemistry, because many people in different countries were doing similar work at the same time and not all of them kept records of their work (or, if they did, they didn't survive).

However, there is a strong case to be made that Antoine Lavoisier, a Frenchman who lived in the second half of the 1700s, deserves the title of 'Father of Modern Chemistry' – not because he invented the science of chemistry, but because of the *methodical* way in which he carried out his experiments and for what he found out from doing them, as described in his book *Traité Élémentaire de Chimie* (published as *Elements of Chemistry* in English, which may or may not have been an intended pun).

The book helped to spread Lavoisier's findings, as well as his ideas about how to conduct experiments and what apparatus to use. This is a key part of the scientific process – scientific ideas and experimental results need to be written down, not only so that they can be passed on to other people, but so that they can be evaluated by other scientists.

The school science textbooks which I read as a student gave the impression that nearly all the great discoveries of science came from white European men like Lavoisier, without really pointing out that women, people of colour and poor people were not generally in a position to have their own laboratories or to spend time doing experiments to satisfy their own curiosity. More recent textbooks make some effort to mention the work of women and non-Europeans whose work, knowledge and records helped lay the foundations of many of the key developments in science.

In Lavoisier's case, his wife, Marie-Anne Paulze Lavoisier, helped him with his work in the lab and kept records of what he did. She also translated books and essays from English into French, so her husband could read them, and drew many of the diagrams for his book *Elements of Chemistry*. There is no doubt that Marie-Anne Lavoisier worked closely with her husband and made significant contributions to his work – it is arguable, as some people have suggested, that she deserves the title 'Mother of Modern Chemistry'.

There is a striking painting of the Lavoisiers in the Metropolitan Museum of Art in New York, which depicts Antoine sat at a table gazing up at Marie-Anne, who is standing. This sort of pose was apparently atypical in portraits of married people at the time and, I like to think, perhaps indicates an acknowledgement of Marie-Anne's contribution to Antoine's work.

Like a lot of the early scientists in history, Lavoisier was a rich man who had the time and money to pursue his interests in science. He designed and commissioned the building of sophisticated new apparatus to carry out his experiments, spending huge amounts of money in his pursuit of ever more precise and accurate knowledge about the behaviour of chemicals. This is something that continues in science today – scientists have ideas about what they would like to observe or measure, and then they have to design and build the equipment that allows them to do it.

Perhaps the most famous modern example of this is the Large Hadron Collider, the huge machine that cost billions of pounds to build and was designed to look for a subatomic particle called the 'Higgs Boson'. Thankfully, unlike in Lavoisier's time, you don't need to be a rich individual to be involved in designing and commissioning specialist equipment, you can work for a university or other research organisation which will pay for it (with money that often comes from the government or charities).

Lavoisier's wealth was inherited from his mother, and he also made money from his involvement in a group that collected taxes for the French government – something for which he would be guillotined in 1794, during the period known as the 'Reign of Terror' of the French Revolution. None of this diminishes the accomplishments of Lavoisier, whose discoveries paved the way for the development of a *scientific* theory that used atoms to explain chemical behaviour. One of the key things Lavoisier did was to show that there were some substances which could not be broken down into simpler substances by chemical reactions.

This supported the relatively new idea at the time of *chemical elements*, and helped to replace the Greek idea of everything being made from the four 'elements' of Earth, Water, Air and Fire. By making very careful, precise measurements, using his expensive, specially made apparatus, Lavoisier also demonstrated one of the most important findings in chemistry, the law of conservation of mass:

> *the total mass of the substances before a chemical reaction is the same as the total mass of the substances produced after the reaction.*

This is taught as *La Loi de Lavoisier* in France, and is not easy to prove for some reactions where one of the products is a gas that can escape into the air.

The Elements

The existence of chemical elements and the law of conservation of mass were two of the *observations* that led the English scientist John Dalton to propose that chemical reactions could be *explained* using a scientific model based on atoms. In the early 1800s, only a few years after Lavoisier's execution, Dalton suggested that *chemical elements are substances which are made up of only one type of atom.* In his model, all the atoms of a particular element were identical to each other, but the atoms of different elements had dissimilar sizes and masses. Like Democritus's ideas about atoms, this helped to explain the distinct properties of different substances, but unlike Democritus, Dalton had based his ideas on the results of his and other scientists' experiments.

So, for example, Dalton's model could *explain* the law of conservation of mass as a consequence of the fact that *no atoms are lost or made in a chemical reaction, they are simply re-arranged.* If you remember having to 'balance' equations for chemical reactions at school, this is why the number of atoms of each element needs to be the same on both sides of the equation. I don't think my chemistry teacher colleagues would mind if I said that this is what their subject boils down to: the study of how atoms become rearranged to make different substances. Even though there are fewer than a hundred different naturally occurring atoms, there are millions and millions of ways in which they can be chemically combined, so there is plenty of work for chemists to do.

The earliest substances identified as elements included silver, gold and many other metals, as well as non-metals such as sulphur and oxygen. Lavoisier showed that water was not an element as the Greeks had thought; instead, it is a chemical *compound*, a substance made from two or more elements chemically joined together. Compounds are formed by the chemical reactions between elements, and they can be separated back into elements using chemical reactions.

One thing that immediately becomes clear from carrying out even a few chemical reactions is that *a compound can have drastically different properties to the elements from which it is made* – sodium chloride (table salt) is a hard, white crystalline substance that is so unreactive and harmless that we can handle it with our fingers and eat it, but sodium is a soft, silvery white metal that is so reactive in air that it has to be stored under oil to prevent it bursting into flames, and chlorine is a strong-smelling, yellow-green gas that can cause skin and eye irritation and is toxic if inhaled. Similarly, none of the main elements in ice cream – carbon, hydrogen, oxygen and nitrogen – are remotely like ice cream when they are not combined to make compounds like sugar and fats.

Experiments show that any particular compound will always have the same *proportion* of different elements in it; for example, if you separate any amount of water, H_2O, into hydrogen and oxygen, you will always get about one-tenth (11.1 per cent) of the original mass as hydrogen and about nine-tenths (88.9 per cent) as oxygen. John Dalton suggested that this could be explained using his atomic theory – the atoms of different elements, being indivisible objects, must always combine in whole numbers to form compounds, a bit like if you make something out of Lego, you always use a whole number of pieces. In the case of water, for instance, two atoms of hydrogen combine with one atom of oxygen to make a water molecule.

You cannot have half an atom or two-thirds of an atom. Again, this is an important idea taught in school chemistry:

A compound is a substance whose basic particles are molecules, made of the atoms of two or more different elements joined together.

Some compounds, like carbon monoxide, CO, are made up from two different types of atom joined in equal numbers. Carbon dioxide, CO_2, is made up of the same two elements, but has molecules made of two oxygen atoms joined to one carbon atom. Other compounds can be made up of many more elements, like the blood pressure medication sodium nitroprusside, $Na_2[Fe(CN)_5NO]$, which has molecules made up of 15 atoms from 5 different elements, sodium (Na), iron (Fe), carbon (C), nitrogen (N) and oxygen (O).

Remembering and writing down the names of chemical compounds and their formulas is one of the things that some students find most challenging about school chemistry (I still take a weird pride in knowing how to spell 2, 4, dinitrophenylhydrazine), but it is a crucial part of studying substances *scientifically*. This wasn't always the case – whenever people discovered new substances, they gave them names, often based on how they looked or some other key property, but sometimes with no obvious connection to the substance at all.

What we call nitric acid today was known as *aqua fortis*, meaning 'strong water' in Latin, and the chemical silver nitrate was known as *lunar caustic*, because it burned flesh on contact (caustic) and silver was associated with the Moon (*luna* in Latin). Substances were sometimes known by different names by different people, and if you didn't know what 'red precipitate' or 'oil of vitriol' was, the name would not be terribly useful. Once again, Lavoisier played a part in changing this, by helping to introducing a 'nomenclature', a *systematic* way of naming chemicals, so that information about the chemical composition of a substance was conveyed by its name.

This is why we call chemicals by names such as arsenic trichloride (which tells us the substance is a compound of one arsenic atom joined to three chlorine atoms) instead of its original name 'butter of arsenic', which makes this poisonous substance sound like something we could eat. Lavoisier said that 'we cannot improve the language of any science without at the same time improving the science itself', and he has been proved to be correct.

A lot of the early chemists, like Lavoisier, spent a great deal of time and effort trying to find or make *pure* substances. In everyday language, 'pure' means 'not mixed with other substances' – so if you buy pure apple juice you would expect to get a container filled with nothing but the juice from apples. However, pure apple juice consists of lots of different chemicals including water, acids and a number of different sugars, so it would not be considered a pure substance by a chemist.

In chemistry, pure substances are ones that consist of only one element or compound, and something like apple juice, which is made of different substances that are mixed together but not chemically joined, is known as a 'mixture'. Ice cream is also a mixture (of mostly water, fat, protein, sugar and air). Most naturally occurring substances on Earth – air, water, soil, wood, rocks, and so on – are either mixtures or compounds.

One reason why it took a long time to identify the hundred or so naturally occurring chemical elements is that only a few of them can be found in their pure form; most have had to be extracted from compounds or mixtures. As scientists discovered more and more elements, they tried to group and classify them based on their properties. Lavoisier was one of the first to group the elements he knew about into metals and non-metals, but it would be more than fifty years after his death when this scientific urge to group and classify things resulted in one of the greatest breakthroughs in chemistry.

School science labs are often decorated with posters depicting 'careers in science' or famous scientists, but these change with the times. For example, I hope you'd see more posters today that feature women and non-white scientists than I did when I was at school. However, there is one constant to be found on the walls of most school science labs: a copy of the periodic table of the elements. There are lots of versions of the periodic table out there – some printed in colour, some in black and white, some with additional information about individual elements, others without.

Whichever version you had in your school, it will have had the same iconic shape and layout, but this was not the shape or layout of what is regarded as the original periodic table, published by the Russian chemist Dmitri Mendeleev in 1869. Other tables listing the elements had been published, but the one we use today is descended from Mendeleev's.

Dmitri Mendeleev was writing a chemistry textbook and set himself the task of organising the elements into a table. He had information about the properties of the 63 elements known at the time, including their 'atomic weights' (the mass of an atom of an element expressed as a multiple of the mass of the lightest atom, hydrogen) and noticed that the properties of the elements were 'periodic' – elements with similar properties seemed to occur at repeated intervals if they were arranged in order of increasing atomic weight.

He put the elements in a table with six columns, so that elements with similar chemical properties appeared in the same horizontal row (in the modern periodic table this has been turned through 90 degrees so that elements with similar properties appear in the same vertical column, known as a 'group').

The thing which suggests that Mendeleev had a rather brilliant mind is not just that he noticed this pattern, but that he also saw it was something *fundamental* in nature. This gave him the confidence to put some elements in a position other than that suggested by their accepted mass, and to leave gaps in his table for elements that had not yet been discovered. He even *predicted* the atomic masses and the physical and chemical properties of these *hypothetical* elements by looking at the properties of the elements next to them in his table.

Imagine his satisfaction when the element scandium was discovered in 1879 and then germanium in 1886, both of which fitted into gaps in his table and had the properties he had predicted. This was convincing evidence that Mendeleev was correct and, within twenty years of the publication of the first textbook in which it appeared, the periodic table was widely accepted as an incredibly useful way to organise knowledge about the elements. The fact that the position of an element on the periodic table allowed chemists to predict its physical properties, and how it would behave in chemical reactions, helped revolutionise how chemistry was done.

According to the chemistry writer and educator Dennis Henry Rouvray, 'Chemistry without the periodic table is as hard to imagine as sailing without a compass'. There is no doubt the periodic table makes it easier to *learn* chemistry, because you don't need to learn about all the elements and their compounds; you only need to know about the groups.

No one is likely to dispute the fact that Mendeleev deserves his place amongst the great scientists of history for coming up with his periodic table, but it's worth pointing out that other chemists at the time, like Julius Lothar Meyer in Germany, had also spotted that the properties of the elements were 'periodic' if they were arranged in the order of their atomic weights. Mayer came up with his own periodic table, independently, and also predicted the existence of elements that had not yet been discovered.

As I've pointed out before, this is an interesting way in which important scientific discoveries are different from great works of art, or music, or literature – we can make the case that no one but Van Gogh could have painted *Starry Night*, or that only Tagore could have written *Gitanjali*, but it's often the case in science that the great discoveries could, and would, have been made by someone other than the people usually credited in school textbooks.

In one sense, the periodic table is simply a convenient way of presenting information, a 'data sheet' given to students for answering questions without them having to remember all the names and properties of the elements. However, putting up a periodic table in a school lab serves another purpose: just as a cross on the wall of a church is a symbol of the Christian faith, derived from the story of Jesus' crucifixion, the periodic table is a symbol of *science* and the tremendous power it gives us in attempting to understand the natural world.

The science of chemistry has not only provided us with an understanding of matter that would have delighted and astonished Democritus, it has given us the tools to create new substances to make our lives better, from many of the medicines and vaccines which we take for granted, to the soft serve ice cream that is pumped out of the machine in an ice-cream van.

Chapter 4
What Is the
Smallest Thing?

'Small is beautiful' is a phrase I've heard a lot . . . mostly from people trying to make me feel better about being an adult who is only 160 centimetres (5 feet, 3 inches) tall. There is indeed something pleasing about miniature objects, as anyone who has played with a really good doll's house will tell you, but I was surprised to learn that the expression comes from the title of a book by the economist E.F. Schumacher. The book puts forward the argument that smaller economies, businesses and cities would be better for human happiness.

Schumacher wrote his book in 1973, and suggested that the population of a city ought not to rise above 500,000. He died in 1977 and would probably be disappointed to know that there are many, many cities in the world with more than ten times the maximum number of people he suggested – my home city of London has a population of nearly 9 million and there are several cities in the world, including Tokyo, Delhi and São Paulo, with populations of more than 20 million.

It seems obvious, but we need *numbers* to talk about the size of a city in this way. Numbers allow us to define a 'small city' as one with 500,000 or fewer inhabitants and a 'megacity' as one with more than 10 million. We also need numbers if we want to talk meaningfully about the mind-boggling range of sizes of objects in the universe.

Most of us take the existence of numbers for granted, but none of us is born knowing what the word for each number means. My youngest daughter, who is a little over two years old as I write this, has learned how to recite the numbers in order up to ten, but it wouldn't be correct to say she can count. She has no real idea what seven means, or that eight is one more than seven, but she can tell the difference between a very big pile of chocolate buttons and a very small one.

In some ways, my daughter is similar to early humans, who didn't have words for numbers but instead were limited to describing things in quantities of more than one or two using the equivalent of 'a few' or 'lots'. The development of languages with numbers was a major step towards a *scientific* understanding of the world, because numbers allow us to *quantify* things and convey information about the world more *precisely*.

I started writing this book in 2020, during the lockdown imposed in the UK to restrict the spread of the COVID-19 virus. A quick look at the news revealed how difficult it would be to communicate about the virus without numbers – there were daily reports of the number of people known to be infected with the disease, as well as the number of people who had died. Most importantly, there was the '(effective) reproduction number', R, also known as R_e, which told us the average number of people that one person with the disease would be likely to infect.

If R was more than 1, the disease would spread to more and more people. The bigger R was, the quicker it would spread. The lockdown, and all the social distancing measures, were aimed at getting this number as low as possible, ideally far below 1, so that the disease infected fewer and fewer people.

The spread of a disease like COVID-19 is only one of the many aspects of nature which can be described and explained using numbers. This is why scientists often talk of mathematics as being the 'language of science' (some have even gone as far as calling it the 'language of God'). Unfortunately, one of the things that puts some students off science at school is that it often seems to involve a lot of maths. But don't worry if you're one of these people – the most difficult bit of maths you'll need for this chapter is knowing how to read the way scientists write very big and very small numbers, and I'll explain that.

If I say something is as small as the width of a human hair, or as big as an elephant, you can probably imagine how big or small it is. Comparisons such as this might have been sufficient for early humans if they wanted to describe the size of things in their immediate environment, but once we want to talk about galaxies or atoms, these kinds of comparisons are not much use.

Measurement

To answer the question 'What is the smallest thing?', we need the concept of *measurement*. A measurement tells us about a *physical property* of a thing – how heavy it is, or how long or how hot, and so on. Numbers on their own are not quite enough for making measurements – it doesn't make sense to say that a bag of rice has a mass of 5 or that a horse is 1.8 tall or that a day lasts 24. To convey *useful* information, we also need *units*, so in these examples we would say a mass of 5 *kilograms*, a height of 1.8 *metres* and a time of 24 *hours*.

A unit is a standard thing to which we can compare other things. These standard things have to be agreed upon by the people sharing information so that if, for example, we are using cubits to measure the length of things, we must all agree how long a cubit is. In the past, this was often achieved by having a physical object that defined the unit; for example, a stick of a certain length could be used to define the cubit, and used to make other sticks of the same length to be used as rulers.

Units for time were usually based on the length of a day, or a year, or the lunar cycle. People in different parts of the world, and even in different parts of the same country, used different units for measuring things, but as people travelled more widely, and traded goods and information with people further afield, it became clear that there was a need for internationally agreed standards for units.

In the 1790s, the French government decided to do something about this and, with the help of scientists from the French Academy of Sciences (*Académie des Sciences*), they produced units of measurement that were not defined by physical prototypes, but based on natural phenomena.

They chose the metre as the unit for measuring length, and defined it as one ten-millionth of the distance on the Earth's surface from the north pole to the equator, on a line passing through Paris. To allow everyone to make practical use of this definition, a bar of this length was made from platinum and kept in the French national archives. The French Academy of Sciences also produced a platinum cylinder to define the unit of mass, the kilogram, which was based on the mass of 1 litre of water at a temperature of 4°C.

Today, nearly every country in the world uses the International System of Units, known as 'SI units', from the abbreviation of its name in French, *Le Système international d'unités*. These are the units we learn about in school science, and include the metre (m) for length, the kilogram (kg) for mass, and the second (s) for time. The precise definitions we use now are no longer based on those the French came up with in the 1790s, but instead make use of physical constants such as the speed of light – the metre, for example, is now defined as the distance travelled by a ray of light in a vacuum in a time of 1/299,792,458th of a second.

Although all scientists in the world use SI units, people in some countries, such as the USA, continue to use other units as well. This had disastrous consequences in September 1999, when NASA's Mars Orbiter, a robotic spacecraft costing over $100m, crashed into the surface of Mars because two different teams working on the project had not used the same units for measuring the force from the thrusters.

Metres, kilograms and seconds are units to which we can relate, because they are on a *human scale* – most grown adults are between 1.5 and 1.75 metres tall, a standard bag of sugar has a mass of 1 kilogram, and we can time seconds by saying something like 'one Mississippi, two Mississippi, three Mississippi' and so on at a steady rate. But if we want to talk about very small things like viruses, or very big things like the Sun, the numbers we use can end up being difficult to handle. The mass of the Sun is about 2,000,000,000, 000,000,000,000,000,000,000 kg. To make big numbers like this more manageable, we can express them in terms of the number of times you have to multiply a smaller number by ten.

For example, two thousand (2,000) can be written as 2×10^3, which is a way of writing two times ten, times ten, times ten. Similarly, twenty thousand (20,000) can be written as 2×10^4, two hundred thousand (200,000) as 2×10^5, and so on. If you did not already know, I hope you can see that the little number above the 10, known as the 'power of ten', is the number of times you multiply the first number by ten. Using this notation (called 'standard form' in the UK) we can write the mass of the Sun as 2×10^{30} kg. Similarly, we can write very small numbers in terms of the number of times we have to *divide* a number by ten.

If we want to describe something that is two-thousandths of a metre long, for instance, we can write it as 2×10^{-3} metres. The *negative* power of ten means you *divide* by ten that many times. So, instead of writing that the length of a virus is 0.00000002 metres, we can write it as 2×10^{-8} metres. The powers of ten used when expressing numbers like this are known as 'orders of magnitude', and are a useful way to *compare* the size of numbers – for example, 10^8 is two orders of magnitude bigger than 10^6, and 10^{-9} is three orders of magnitude smaller than 10^{-6}.

To make things more convenient, some of these multiples of ten are given their own name to be used as prefixes with units, for example the word 'kilo' in front of a unit means it is multiplied by a thousand (a *kilo*metre is 1,000 metres), the word 'micro' means the unit is divided by a million (a *micro*metre is a millionth of a metre), and the word 'nano' means the unit is divided by a billion (a *nano*metre is a billionth, 1/1,000,000,000th of a metre).

Numbers and units allow us to count and measure things, but when it comes to very big or very small things, they do more than that – they allow us to conceive, and make sense of, things that we would otherwise not be able to do. An atom is unimaginably tiny – we literally cannot imagine how small it is. One way to convey the size of an atom is to say something like 'five million atoms could fit across the width of the full stop at the end of this sentence'. We can say this, because we know that an atom is about 10^{-10} metres wide and a full stop on this page is about half a millimetre (5×10^{-4} metres) across, and dividing 5×10^{-4} by 10^{-10} gives us 5 million.

Although this doesn't actually help you imagine the size of an atom, because you can't picture five million things side by side, the fact that we can put numbers to statements like this gives us a way to communicate about their size even though we can't visualise them in our minds. Without numbers, we wouldn't really have a way to get our heads around just how big or how small the biggest and smallest things are.

One of the problems with dealing with a virus such as COVID-19 is that it is invisible – we can't see it with our eyes because it's so small, so we don't know where it is. This is why we need to use special hand-washing techniques – we can tell if we have washed dirt or grease off our hands just by looking, but to remove any traces of a virus off our hands, we need to make sure we cover every bit of them with soap.

A virus cannot even be seen with an optical (light-based) microscope of the kind you should have used at school. I say 'should' have used at school because microscopes are remarkable instruments, which allow us to observe things that otherwise would be *impossible* to see. Looking through one is something that everyone ought to have the opportunity to do. The smallest things humans can see with our unaided eyes are about the width of a human hair, or 0.1 mm wide, but with the help of a microscope we can see things that are hundreds of times smaller.

This in itself is enough to amaze most children, but there is a deeper lesson to be taught: the first microscopes allowed people to see things that they had not even *imagined* existed, and dramatically changed their ideas about the nature of the world, and in particular the nature of living things.

Microscopes, and the telescopes that were invented around the same time, are examples of *scientific instruments*, tools which extend our ability to study nature beyond the limits of our human senses. It's unlikely science would have progressed very far if people had not invented instruments – just imagine how different our understanding of the world would be without thermometers to measure temperature, or weighing scales to measure mass, let alone without DNA sequencers and particle accelerators.

I can still remember the first time I used a microscope, nearly forty years ago, looking at a sample I had taken from inside my cheek and another from the thin skin I had carefully peeled from an onion. For some reason, the experience is one that has stayed with me – I can recollect scraping the inside of my mouth with a cotton swab, then smearing the end of the swab across a glass microscope slide and adding a drop of blue dye, before fiddling with the various knobs on the microscope to get it into focus.

I think the memory might have stuck because it seemed so astonishing to me that I was looking at *my own cells*, that these pale blue blobs with the dark blue 'nucleus' were not just some random cells, like the photos in the biology textbook, but something that was uniquely a part of *me*. I suspect the sense of awe and wonder also came from realising that these things which I had been learning about in my biology lessons really were there for me, or anyone else, to see.

This is another reason why I think it's so important for *every* child to have the experience of using a microscope as part of their education: the act of looking at cells for yourself, and not at photographs or diagrams, means you don't have to take your science teacher's word for it that cells exist, it means you get to *verify* for yourself that cells are real, concrete things, that plant cells are different to animal cells, and that they have the features you've been told you'll need to know for the end of term exam.

Verification of what people *claim* about how the world works is such an important aspect of science that the motto of the UK's Royal Society, the world's oldest scientific institution, is *Nullius in verba*, the Latin for 'take nobody's word for it'. The founders of the Royal Society were clear that scientists should not take things on authority, but check facts for themselves. Of course, it's not practical for all of us to scrutinise every scientific fact for ourselves, but in principle all scientific findings ought to be verifiable by anyone with enough time and resources.

There may have been another meaning to the motto that the founders of the Royal Society wanted to convey, one that is more political in its nature: science should not *automatically defer* to those in authority, and social status and power should be irrelevant when it comes to knowledge about how the world works.

Hooke's Law

One of the laws of physics that nearly all school children test for themselves is Hooke's Law, which says that the extension of a spring is proportional to the force pulling it. By hanging weights one by one from a spring attached to a clamp, and measuring the increase in its length, students usually end up confirming that Robert Hooke was right – if you double the force, you double the extension, if you triple the force, you triple the extension, and so on (unless you use too many weights and the spring becomes permanently stretched or breaks). It's not the most exciting of experiments and the results don't really surprise anyone, but it does show that the world behaves in the way physics says it ought to.

Hooke's Law was all I learned about Robert Hooke in school and it wasn't until much later that I found out that he was something of a *polymath*, with knowledge and expertise in many diverse areas of science, ranging from astronomy to geology and physiology, as well as being a surveyor and architect who played a key role in rebuilding London after the Great Fire of 1666.

If you have a car, the universal joint that transmits the rotation of the engine to the rotation of the wheels is based on a design by Hooke. In a lecture he gave about Hooke's life at the Royal Society in 1949, the English physicist and science broadcaster Edward Neville Da Costa Andrade said that Hooke was 'probably the most inventive man who ever lived, and one of the ablest experimenters'.

Explaining why Hooke was not as widely remembered as other great scientists of the time, Andrade said that 'Hooke made many discoveries of capital importance', but 'he would come upon some brilliant and original mechanical conception, take it a certain distance, generally far enough to show that it would work, and then, stopping short of the final stage and precise publication that would convince the world of his priority, take up some other brilliant notion that had come into his head.'

Hooke's life and work are fascinating, revealing how messy and *human* the history of science is, unlike the neat and tidy version that is often presented to school students, where an individual scientist has an idea, does an experiment, and comes up with a law.

One aspect of Robert Hooke's work that was very much appreciated during his lifetime, and continues to be, was his book *Micrographia: or Some Physiological Descriptions of Minute Bodies Made by Magnifying Glasses. With Observations and Inquiries Thereupon.* Published in 1665, the book was filled with Hooke's beautiful, detailed drawings of everyday things such as snowflakes, insects and razors, as seen through a microscope. A previously hidden world was unveiled in these illustrations, capturing the public imagination and leading the book to become what some people have described as the 'world's first scientific best-seller'.

Samuel Pepys, now famous in his own right for the diaries he kept at the time, called it 'the most ingenious book that I ever read in my life'. Pepys ended up so interested in microscopy that he bought his own microscope at a 'great price'. However, like many a school student using a microscope for the first time, Pepys found it was not straightforward to use the instrument or to see exactly what Hooke had seen. Part of what made Hooke exceptional as a scientist was his skill and perseverance as an experimenter. He went to great lengths to get the images he wanted, for example dunking insects in brandy to make them sufficiently docile while under his microscope, and spending entire days at a time to study just one sample.

Hooke did not simply draw what he saw, instead he composed his illustrations from many different observations of the same thing, using different lenses and lighting conditions. A quick image search on the internet for 'head of a housefly' will, as you might expect, give you lots of close-up photographs of the head of a housefly. Many of these images are taken using computerised imaging techniques which involve combining several images taken by focusing the camera at different distances. Hooke did something like this without the aid of computers, and it is striking to see how similar his drawing of a fly's eye is to modern images made with much more sophisticated equipment.

Hooke's images revealed intricate details of the unseen world and showed that things on the microscopic scale were as complex as ones in our everyday world. One of the lasting legacies of Hooke's work with microscopes is the word 'cell', which we use to describe the basic building blocks that make up animals and plants – Hooke was the first to use the word to describe the microscopic structures he saw when he studied cork (the bark of the cork oak tree).

Hooke was not the only scientist at the time who was using microscopy to discover new and exciting things about the world. In Italy, Marcello Malpighi's microscopic studies of human flesh revealed the existence of the tiny blood vessels we call 'capillaries', and, over in the Netherlands, Antony van Leeuwenhoek studied rainwater using his own design of a single-lensed microscope to study water from various sources, becoming (probably) the first person to see the smallest *living* things anyone had ever seen: microorganisms and bacteria.

Micrographia demonstrated to the public, and to other scientists, the microscope's huge potential as an instrument for discovering new knowledge about the world. If he was alive, Hooke would not be surprised to learn that, over 350 years after the publication of his book, microscopes still play a key role in scientific research and discoveries made with them have helped us understand everything from the processes that drive life to the formation of rocks.

Hooke wrote that 'by the help of Microscopes, there is nothing so small, as to escape our inquiry', but there is a limit to the smallest thing that can be seen with even the best microscope. To understand why, we need to go back to how we see things in the first place.

In the chapter 'Why Is the Sky Blue?', I told you that we can explain some of the behaviour of light using the *ray model of light*, in which we use imaginary lines to show the path that light takes. We can use this model to explain how we see an object: light rays travel from the object into our eyes and form an *image* where they hit the retina. Our brain translates this into what we see in our minds. The *size* that we see an object depends on how much space the image takes up on our retinas, and this depends on the angle at which the rays enter our eyes. If you move the object closer, the angle widens so as to make the object look bigger.

We can achieve the same effect by looking at objects through a magnifying glass, because rays of light can change direction when they pass through glass and into the air, an effect called 'refraction'. Microscopes work using the same principle, but can produce larger images than a single lens by making use of two or more lenses. The *magnification* of a microscope tells us how much bigger the image of an object is compared to its actual size. The maximum *useful* magnification that can be achieved with a microscope is about 1500. If we magnify things more than this, we don't see any more detail, all we get are blurry images.

Once again we can use a model of light to understand why this is the case, but this time we need to use the *wave model of light*, which says that light travels as waves and that different colours of light have different wavelengths (the distance from one peak of a wave to the next). Visible light, the light our eyes can detect, has wavelengths of a bit less than a millionth of a metre (or a thousandth of a millimetre). We can see objects that are bigger than these wavelengths with either our naked eyes or with microscopes, because the light bounces off in straight lines and behaves as the ray model of light says it should.

However, when it comes to objects that are around the same size or smaller than its wavelength, light does not travel in straight lines, but instead bends around them, an effect called 'diffraction'. This is how all waves behave – you can see waves on water diffracting around objects like boats in their way. In the case of light, this bending means that instead of bouncing off an object in straight lines and forming a sharp image in our eyes, the light spreads out so that we see nothing but a blur. This effect is also why streetlights and the Moon have a blurry ring or halo around them on a foggy or misty night – the light from them is diffracted by tiny water droplets in the air.

Finding the Size of an Atom

I said earlier that the COVID-19 virus is too small to be seen using the kind of microscopes I've been talking about. And yet, many reports of it in the media featured images of the virus, all of which looked something like a sphere covered in knobs or spikes. COVID-19 is an abbreviation for Coronavirus Disease 2019, and the word 'coronavirus' comes from the Latin for 'crown', which is what the scientists who named this type of virus thought it resembled. The images of the virus were made using an *electron* microscope, which makes use of a beam of electrons instead of visible light. In simple terms, the electrons (I'll explain exactly what they are shortly) are fired at the object being studied, and bounce off into a detector connected to a computer, which processes the information to produce an image.

An electron microscope can show us far smaller things than an optical microscope, because electrons used in this way *behave like waves* but with a wavelength thousands of times smaller than that of visible light. The images of COVID-19 in the media are not all of the same colour – in some that I've seen, the virus is grey and red, in others it is blue and purple. This is because things that are smaller than the wavelength of visible light are not 'coloured' and the images from electron microscopes are black and white. The colours are added by scientists to highlight the different features of the virus.

Microscopes played a significant role in extending our knowledge of very small things, but it wasn't through microscopy that the smallest things known to science were discovered. As I explained in the chapter 'Why Does Ice Cream Melt?', people had guessed that matter was made of tiny, invisible particles long before the invention of the first microscopes. By the early 1900s, atoms were more than just an idea, they were regarded as being real, even though they could not be seen. Scientists did not need to *see* atoms to accept that they existed, as there was plenty of other evidence. For a while, the answer to the question 'What is the smallest thing?' was the hydrogen atom, which chemists had shown was the lightest of the different types of atoms.

So, how do we know the size of an atom? Obviously, you can't use a ruler or similar instrument to measure one *directly*, but there is a remarkably simple way to get a good *estimate* for the size of an atom, which I was lucky enough to do at school. I say I was lucky, because it's not something that is done widely in schools these days, but I think it's a brilliant way for children to experience for themselves how we can *deduce* information about things we can't see, using knowledge and measurements about things we can.

The method, which makes use of the fact that oil floats on water, was devised in the late 1800s by the English physicist John William Strutt, usually known as Lord Rayleigh, because he was a baron. The way we did it in school was pretty much the same way Rayleigh did it: we filled a large tray with water and sprinkled a very fine white powder onto the surface. We then dipped a small loop of wire into some olive oil to capture a tiny drop of it, and measured its diameter using a magnifying glass.

The drop of oil was less than a millimetre wide and we placed it gently onto the surface of the water, where it spread out. The oil left a clearly visible, roughly circular, dark patch where it had pushed aside the white powder we had sprinkled on the water. By measuring the diameter of this patch of oil, we had collected the information we needed to calculate the approximate size of an atom.

All we needed to do now was some maths, which I'll outline below (you can skip to the end of the next paragraph if you just want to know how big an atom is):

The patch of oil has the same volume as the drop, but is in the shape of a very, very flat *cylinder. Because of the way oil behaves, it's reasonable to assume that the cylinder of oil floating on the water is only one molecule thick. The volume of the original drop of oil can be calculated using the formula for the volume of a sphere, $V=\frac{4}{3}\pi r^3$, where r is the radius (half the diameter) of the drop. This value can then be equated to the formula for the volume of a cylinder, $V=\pi r^2 h$, where this time r is the radius of the patch of oil, to find a value for h, the height (thickness) of the cylinder. This is the approximate length of one molecule of oil. From chemical analysis, it's known that olive oil molecules are made up of a chain of twelve (carbon) atoms, so the length of one atom is the length of one molecule divided by twelve. If this experiment is done carefully, even schoolchildren can show, as Lord Rayleigh did, that the diameter of an atom is about 10^{-10} metres. This is not an exact number, but it is the correct order of magnitude, as confirmed by more sophisticated modern methods.*

For a long time, people had imagined atoms as minuscule, indestructible spheres, but this changed in 1897 thanks to the English physicist J.J. Thomson. Thomson had been investigating the properties of glowing rays, which could be produced by passing an electric current between two metal electrodes in an evacuated glass tube (one from which most of the air had been removed). Known as 'cathode rays', because they came from the electrode called the cathode, these rays could be easily bent with a magnet, so that they followed a curved path instead of travelling in a straight line.

If they had been rays of *light*, they would not have done this, but Thomson knew that *electrically charged* particles could be made to change direction by a magnet. 'Electric charge' or just 'charge' is a property of matter that is responsible for the crackling sound you hear when you take off a jumper made of synthetic fibres, the way your hair stands on end when you whizz down a plastic slide and that shock you sometimes receive when getting out of a car. Charge is also why electricity exists, and students usually learn about it under the topic of 'static electricity'. In most secondary schools, the existence and properties of charge are demonstrated using a big, shiny, domed contraption known as a 'Van de Graaff generator'.

If you missed out on this at school, go and try the following experiment, which is how I introduce the subject when I teach:

Find a plastic straw, or a plastic pen (one of those cheap biros is ideal), and some kind of drink bottle with a lid. You'll also need a cotton rag or T-shirt – the one you're wearing will do. Once you've got these items, do this:

- If you're using a pen, remove the lid and any other parts you can, so that you are left with just the plastic casing.
- Grip the straw or pen with your T-shirt and give it a good rub up and down, as if you were trying to polish it.
- Balance the straw or pen sideways on top of the bottle.
- Now, using all your mental powers, *will* the straw or pen to move towards you while you bring your hand close to one end of the object (but don't touch it). Imagine that you are a Jedi Knight, if this helps.

If you've done this correctly, and haven't seen it before, you've probably stopped reading and are now running around trying to find someone else to show this amazing feat of telekinesis – it really does look like 'magic'. What should have happened is that the straw or pen moved towards your hand, a bit like when you move a magnet close to a metal object or another magnet. If, instead of your hand, you use another rubbed straw or pen, the first straw or pen will move away instead of being attracted. The straw or pen will also attract small scraps of paper and, perhaps most impressively, if you hold it near a thin stream of water running from a tap, you can make the water bend towards you in another disconcerting display of this phenomenon.

These are displays of 'static electricity', a phenomenon that people have known about for thousands of years, although they didn't always call it that. The words 'electric' and 'electricity' are derived from the Greek word for amber – *elektron* – because Ancient Greek philosophers had seen that objects made of this material were particularly good at attracting things like feathers and dust.

Arguably all science comes from people who take an interest in the natural world, people who see something in nature that fascinates them and want to find out how it works. Static electricity is something that intrigued lots of people. Amongst those who studied it are the British philosopher Francis Bacon, and Benjamin Franklin, whose face appears on the United States' $100 bill.

In 1752, Franklin famously flew a kite in a thunderstorm to show that lightning was a form of static electricity. The kite had a metal wire at the top and a wet string at the bottom, onto which a metal key was attached. Franklin knew that if there was any static electricity in the sky, it would flow through the metal wire and the wet string. If there was enough of it, he would be able to feel an electric shock when he touched the key. The experiment worked exactly as Franklin thought it would, and he went on to invent lightning conductors to keep buildings safe from lightning strikes.

Franklin, and others at the time, knew that certain objects could be 'electrically charged' by rubbing them. They also knew from their experiments that there were two types of charged object, which they called 'positive' and 'negative', and that similarly charged objects repelled each other while oppositely charged ones attracted. But until 1897, scientists did not have a good explanation for what *actually* happens when you charge up an object by rubbing it. Thomson's experiments with cathode rays would help change that.

Smaller than an Atom

Thomson knew that the exact effect a magnet had on a moving charged particle depended on how much and what type of charge it had (positive or negative), its mass, and how quickly it was moving. Other scientists at the time suspected that cathode rays were made up of negatively charged particles, but Thomson's experiments, measurements and calculations showed that these mysterious glowing beams consisted of vast numbers of *identical* charged particles, all moving at the same speed, and with the same charge and mass. If the particles were not identical, the rays would have become wider or split up when they were deflected by a magnet.

Most significantly, Thomson's calculations showed that these particles had a mass of less than a thousandth of the mass of the smallest atom, hydrogen. It soon became clear that he had discovered the first thing smaller than an atom, the first *subatomic* particle: the electron.

Thomson repeated his experiments using glass tubes filled with different gases, and with electrodes made of different materials, and showed that the cathode rays were the same in each case. This meant that unlike atoms, which come in different types depending on the element they belong to, all electrons are the same. Today, over a hundred years after it was first discovered, and despite efforts to show otherwise, scientists believe electrons are a truly *fundamental* particle, one that cannot be broken down into anything smaller.

The details of exactly how Thomson proved the existence of electrons are beyond the scope of this book, but what is worth pointing out is that, by analysing the behaviour of something visible, Thomson had discovered the existence of a particle which could not be seen. In this way, his work was similar to how Einstein had shown that Brownian motion (which could be seen) was evidence of the existence of atoms (which could not be seen), and Rayleigh's estimate for the size of an atom. Thomson's discovery relied on a form of reasoning known as *induction*, where new knowledge is arrived at by piecing together existing knowledge, observations and experimental results.

The discovery of the electron led to a new model of the atom. Some scientists suggested that, instead of solid spheres, atoms might be balls of dense, positively charged matter dotted with these tiny, negatively charged electrons. This became known as the 'plum pudding' model, because it resembled a stodgy, steamed sponge pudding with embedded raisins (known as 'plums' at the time), which went by that name. If the model had been invented in modern times, it might have been called the 'chocolate chip muffin' model, although the fluffy texture of a muffin is not quite as good an analogy for the dense, positively charged matter it would represent.

The plum pudding model could be used to explain the fact that atoms were not charged overall – the negative charge from the 'plums' (electrons) are balanced by the positive charge of the sponge, leaving a *neutral* atom. The model could also account for the existence of different elements – different atoms would be made up of varying amounts of sponge material with different numbers of plums. As for phenomena related to static electricity, like the demonstrations with the 'magic' straw described above, the discovery of the electron helped scientists to understand that these effects happen because of the transfer of electrons from one object to another: things become negatively charged if they acquire electrons or positively charged if they lose some.

When you charge something up by rubbing it, for example, you are literally moving electrons from the atoms of one object to the atoms of another. In the case of the straw, rubbing it against the T-shirt causes electrons from the T-shirt to be transferred to the straw, leaving the straw negatively charged and the T-shirt positively charged. The charged straw is then attracted to the positively charged parts of atoms inside your hand, and other objects to which it is brought near. Scientists use the word 'ion' to describe atoms or molecules that have gained or lost one or more electrons, and 'ionisation' to describe processes which lead to this happening.

The plum pudding model had a very short life, and nobody thinks of atoms as being like this anymore. Today, the most widespread image of an atom is one that resembles a tiny 'solar system' with the negative electrons orbiting a positive 'nucleus'. Representations of this might vary from place to place – with different numbers of electrons orbiting in either concentric or overlapping orbits, or showing or not showing the protons and neutrons inside the nucleus – but it's an image that is engrained in our culture, reproduced in countless places, from T-shirts to company logos, to official coins and stamps.

What is less well-known is that the first person to suggest that the negative electrons 'orbited' a positive nucleus was the Japanese physicist Hantaro Nagaoka. He knew the plum pudding model could not be correct because a negative charge could not be 'embedded' in a positive charge, so he put forward his alternative model in 1903. European and American school physics textbooks don't mention Nagaoka's work, but most Japanese ones do.

Whilst his ideas may not be something Western schoolchildren (or even adult scientists) learn about, many prominent European scientists at the time knew Nagaoka and were aware of his model of the atom. One of them was Ernest Rutherford, who devised one of the most famous experiments in the history of science and provided the evidence that would cement the solar system model of the atom in the popular imagination.

Rutherford was a New Zealander who joined J.J. Thomson's team at the University of Cambridge in 1895. It was an exciting time to be an experimental physicist – around the same time that Thomson was carrying out his experiments with cathode rays, scientists elsewhere were investigating mysterious rays of another type, given out by substances containing the element uranium. Unlike other things that emitted radiation – candle flames or electric lights or the kind of apparatus which Thomson had been using – these substances did not need to have anything done to them to release the radiation; they did not need an external trigger or energy source, but were actively emitting it all the time.

This led Marie Curie, the brilliant French-Polish scientist who was studying the phenomenon, to call it 'radioactivity'. Curie and her husband Pierre, along with Henri Becquerel, shared the 1903 Nobel Prize in Physics for their work laying the foundations for our understanding of radioactivity – Becquerel was rewarded for being the first to discover the phenomenon, and the Curies were recognised for showing that uranium was not the only radioactive element, and for discovering two previously unknown elements, polonium and radium, which were both more radioactive than uranium. Sadly, this work was probably the cause of Marie Curie's death – she died from aplastic anaemia, a blood disease that can be caused by exposure to large amounts of this kind of radiation.

In 1899, Rutherford, now working at McGill University in Canada, showed that radioactive substances emitted two different types of radiation, which he named 'alpha' and 'beta' rays. Two years later, a French physicist, Paul Villard, showed there was also a third type, which were named 'gamma' rays (alpha, beta and gamma are the first three letters of the Greek alphabet). Rutherford went on to show that alpha radiation was made up of *particles* and that these particles were positively charged helium ions (helium atoms without their electrons), the second lightest element after hydrogen.

Because these helium ions were being emitted from substances that did not contain helium, it became clear they must be coming from *inside* atoms of the bigger elements the radioactive substance contained. It seemed that radioactivity happened when big atoms spat out smaller ones, changing into a different element as a result. This was astonishing to scientists at the time. Like Thomson's discovery of the electron, it was further evidence that atoms were *not* the fundamental building blocks of matter. It was clear now that atoms were definitely not the smallest thing, and nor were they indestructible, as had been believed for thousands of years.

Earlier, I explained how scientists had worked out that atoms were around 10^{-10} metres wide and also explained why we can't see things of this size using visible light. At the start of the twentieth century, it would have seemed impossible that we could see *inside* an atom, but Rutherford realised that, since alpha particles were smaller than atoms, they could be used to *probe* atoms. He reasoned that, by firing alpha particles at atoms, and seeing how they travelled through or bounced back, it might be possible to learn something new about what atoms were like.

He put this idea to the test in what is widely known as the *Rutherford Alpha Scattering Experiment* or the *Rutherford Gold Foil Experiment*, in which a detector was used to find out what happened to alpha particles when they were fired at a sheet of very thin gold. The alpha particles were produced by a sample of radium, which was placed behind a lead screen with a slit in it, so that only a narrow beam of alpha particles emerged. This, along with the gold target, was placed inside a container from which all the air had been removed, so that the only thing in the path of the alpha particles was the gold.

Gold was used because it could be made into extremely thin sheets, only a few atoms thick. The detector was a screen that produced a tiny, tiny flash of light every time an alpha particle hit it. The experiment is also known as the *Geiger-Marsden Experiment*, because it was actually Hans Geiger (an assistant of Rutherford's) and Ernest Marsden (a student), who carried out the rather tedious work of sitting in a pitch black room staring at the detector for hours on end, waiting to see these 'scintillations'.

If the plum pudding model of the atom was correct, the scientists should have seen all the alpha particles pass through the gold foil, with some being slightly deflected from a straight-line path. And this is exactly what Geiger and Marsden saw . . . except that a tiny number of alpha particles strayed a little off from a straight-line path and, most interestingly, about 1 in 8,000 bounced almost straight back in the direction they had come from. Many textbooks and history books quote Rutherford as being astounded by this, and saying, 'It was as if you fired a 15-inch shell at a sheet of tissue paper and it came back to hit you.'

Some people, like the physics teacher and textbook writer David Sang, have pointed out that Rutherford was perhaps not as surprised as he suggested and that he was probably aware of the possibility of the alpha particles 'back scattering', as the experimental apparatus had been designed so that it *could* detect particles which did this. But whether Rutherford was genuinely surprised or not, his analogy was a good one and helped convey to everyone else how exciting the outcome of this experiment was – alpha particles were indeed like shells (bullets), and the gold foil was like tissue, making the observation that some of them bounced right back difficult to explain.

The Rutherford Alpha Scattering Experiment is not one that students can do in schools, but is one that most of them are expected to learn about. This is not just because of what the experiment revealed about the atom, or because it is regarded as one of the 'great turning points in physics', but because it exemplifies many of the aspects of *how science works*, that is, how *scientific thinking* and *experimentation* can provide information and knowledge about the world in a way that no other approach can.

The Nucleus

The way I remember being taught about this at school, and the way it is presented in most textbooks, is that Rutherford got the results of this experiment and then came up with the solar system model of the atom. In reality, it took him about a year to work out an explanation that made sense of the fact that most of the alpha particles passed straight through the gold foil, but a tiny number bounced back.

Rutherford knew that alpha particles were positively charged, he also knew that atoms were made of both positively and negatively charged components, and that similarly charged objects repel each other. So, he suggested (and used some mathematics to show) that the reason why so few positively charged alpha particles bounced back from the gold foil was because the positive charge belonging to a gold atom must be concentrated in a very small space inside the atom. This was the idea of an atomic *nucleus*, which we now accept as a fact.

Rutherford used the results Geiger and Marsden had collected to show that this positively charged nucleus was around a hundred thousand times smaller than a whole atom, about 10^{-15} metres in diameter. Since electrons could easily be removed from the atom, it made sense to think that they were on the 'outside' of the atom, as Nagaoka had suggested, and along with Rutherford's evidence of a tiny nucleus, a model *resembling* planets orbiting the Sun became widely accepted by scientists.

By 1932, about twenty years after Geiger and Marsden had carried out their experiment, Rutherford had shown that the nucleus contained a positive particle which he named the proton, and a physicist called James Chadwick had shown that the nucleus of all atoms, except for hydrogen atoms, also contained a neutral particle, the neutron. As a result of these discoveries, scientists could confidently state that:

- atoms are made up of three smaller particles: positively charged protons, neutrons with no charge and negatively charged electrons
- atoms are neutral overall because the positive charge of their protons is balanced by always having the same number of negatively charged electrons (an atom which loses or gains electrons is called an *ion*)

• all atoms of the same *element* have the same number of protons, but can have different numbers of neutrons (atoms of different elements have different numbers of protons)

So, if you were some kind of super-powerful cosmic toddler playing with the building blocks of matter, your parents could buy you a kit made up of just under 100 different types of atom, from which you could build everything in the world around us, or they could buy you a kit that consisted of only three different types of piece – electrons, protons and neutrons, from which you could then make any type of atom and therefore any type of substance.

As I've said before, scientists believe that electrons are truly *fundamental* particles, because there is no evidence that they can be broken down into simpler objects. However, the same is not true of protons and neutrons. By the late 1960s, further experiments had shown that protons and neutrons were made up of two different types of *quarks*, known as 'up' and 'down' quarks, with protons containing two up and one down quark, and neutrons having two down and one up quark. Like electrons, these appear (so far) to be truly fundamental particles, because it seems impossible to smash them into anything smaller. So, our ultimate cosmic building-block kit can now be made from electrons, up quarks and down quarks.

Electrons have a minuscule mass – about 1/2000th that of the smallest atom, hydrogen. Most of the actual mass, the 'stuff' that makes an atom, is concentrated in the nucleus, which leads to a conclusion about the nature of atoms that may leave you flabbergasted if you do not already know it: *most of an atom is empty space*. (If you're wondering why, if this is the case, things don't simply pass through each other, the answer is because the electrons on the outside of all objects repel each other incredibly strongly when they come very close to each other.)

It is impossible to draw a diagram in a textbook that correctly shows the relative sizes of the atom and its nucleus. If you wanted to draw a scale diagram with a nucleus being just 1 millimetre across, the whole atom would be 100 metres wide. Lots of people have come up with analogies to get this mind-boggling fact across, including saying that if the atom were the size of a large sports stadium, the nucleus would be the size of a pea.

My favourite analogy is by the playwright Tom Stoppard, who described it like this in his play *Hapgood*:

Now make a fist, and, if your fist is as big as the nucleus of an atom, then the atom is as big as St Paul's, and if it happens to be a hydrogen atom then it has a single electron flitting about like a moth in an empty cathedral, now by the dome, now by the altar.

Stoppard's analogy is not quite right – the nucleus would be much smaller than a human fist if an atom were the size of a cathedral – but his description of an electron 'flitting about like a moth in an empty cathedral, now by the dome, now by the altar' is actually more accurate than the simplistic model of electrons going around in neat circular orbits. The simple solar system model is an over-simplification of what atoms are like, and has several problems. Electrons orbiting a nucleus ought to emit X-rays, which would cause them to lose energy and fall into the nucleus, but since this doesn't happen, it means electrons cannot be orbiting the nucleus in the same way that planets go round the Sun.

In fact, the question of what exactly electrons are like, and what they do inside an atom, comes back to the fact that electrons, and indeed quarks, are not like anything we personally experience in our world, so the approach of using things we already understand to explain things we don't breaks down. We run out of models based on everyday life.

This strange behaviour, sometimes known as 'wave-particle duality', is entirely accepted by scientists, as exemplified by the fact that J.J. Thomson won the Nobel Prize in Physics in 1906 when he showed electrons were particles and his son, G.P. Thomson, won the Nobel Prize in Physics in 1937 for showing that electrons are waves. In order to predict how electrons will behave, or to explain their behaviour in particular situations, we sometimes have to think of them as particles, at other times as waves. This is the same wave-particle duality that we end up talking about when we try to understand the nature of light.

One weird consequence of this is that it doesn't really make sense to ask how 'big' an electron is, because the very concept of size becomes difficult to pin down when we come to measure it for an electron. In some situations, an electron can behave as if it is as big as the atom it is in, but in others, it acts as if it takes up no space at all. Quarks are the same – three of them together make up a proton or a neutron, which are both about 10^{-15} metres wide, but this does not mean that a quark is a third as big. According to our current understanding, quarks are 'point-like' objects, that is, they are infinitely small (although, like electrons, they can sometimes have a 'size', but that size is never bigger than 4.3×10^{-17} metres).

So, what is the smallest thing? The truth is, this is one question that we don't really have a good *consensus* answer for yet. It could be that quarks or electrons are indeed the smallest things, or we might one day discover something else which is even smaller. We don't know. And that's okay. Because not knowing is what drives scientists.

Chapter 5
What Are Stars?

'Baby, I'm a Star' by Prince is one of those songs that instantly transports me back to my school days. The song features in the film *Purple Rain* and perfectly encapsulates the kind of swagger that many teenagers aspire to, but only someone as phenomenally talented as Prince can pull off. If you don't know it, I encourage you to stop reading and find a way to listen to it right now – not only because it is guaranteed to make you feel joy, but because if you pay attention to the lyrics, you'll hear Prince sing something that, for many people, is the most wondrous fact in science: 'We all are a star.'

Okay, maybe he didn't mean it in quite the same way I'm going to tell you about, but just as listening to Prince for the first time gave me goosebumps, so did finding out that the atoms I am made of, the atoms we are *all* made of, were once part of a star. In a very real sense, pretty much *everything* in the world, from this book you're holding to the air you're breathing, is made of stardust. Before getting into exactly what this means, let me quickly list a few other facts about stars that are often described by my students as 'mind-blowing':

- There are more stars in the universe than there are grains of sand on all the beaches of the world (interestingly, there can be more atoms in a grain of sand than there are stars in the universe).

- Our Sun is so big, you could fit over a million planet Earths inside it, and there are stars which are so big that *hundreds of millions* of our Sun would fit inside them.

- The inside of a star is *unimaginably* hot, with temperatures up to hundreds of millions of degrees.

- Stars exist in a *fourth state of matter* known as a *plasma*, made up of positive ions (atoms that have had electrons knocked off of them) and free electrons.

- The light from our Sun takes eight minutes to reach us, and light from some of the stars we can see at night has travelled for thousands of years, meaning that when we look at stars, we do not see them as they are today, but as they were when the light left them, so . . . get ready for this . . . science tells us *we are literally looking into the past when we stargaze.*

This last one is probably a close second, or equal first, in any list of 'amazing facts about stars', but I don't think any of these is the most astonishing thing I teach my students – to me, *the most surprising thing about stars is that we know anything about them at all.*

The Sun is our nearest star. For most of our history, humans did not really know that the big, blindingly bright white disk in the sky, which lights up our world in the daytime, is the same type of object as the tiny, twinkling, pin-pricks of light we see at night time. The Sun is about 150 million kilometres away from us. If we travelled there at a speed of 1,000 kilometres an hour (roughly the fastest a commercial aircraft flies), it would take over seventeen years to get there. Our next nearest star is Proxima Centauri and, travelling at the same speed, it would take *four and a half million* years to reach it.

The distances to stars are so vast that scientists don't use kilometres or miles to measure them; instead, they use a unit based on the time it would take to reach them if we could travel at the fastest speed *possible* – the speed of light, 300 million metres per second. Astronomers would say that Proxima Centauri is 4.24 *light years* away from us, meaning it takes about four years and three months for a beam of light from the star to get here. The term *light year* makes it easy to confuse it for a unit of time, but a light year is a unit of *distance*, with one light year being equal to the distance a beam of light travels in a year.

I hope you can see from this why I think it's amazing that we know *anything* about stars at all – they are almost inconceivably far away. We can't go and physically measure how big they are, or take their temperature, or take a sample for chemical analysis. This is why, for a long time, even scientists believed that it would be *impossible* for us to know what stars are made of, or how they shine. So how *do* we know anything about stars? The answer lies in an experiment you probably did at school, and the fact there *is* one thing that we can get hold of from a star – its light.

In all my years of teaching, I can't recall ever meeting a child who was not excited at the prospect of using a Bunsen burner for the first time. Being trusted to use fire, fragile apparatus made of glass, dangerous chemicals and all the other paraphernalia of a secondary school science lab is a key moment in a child's transition from primary school to the more grown up education that lies ahead. It's usually early on in these new science lessons that children carry out the experiment that helps us understand how scientists know what stars are made of – the flame test, in which you take a clean piece of wire and dip one end of it into a powdered chemical before holding it in the middle of a pale blue Bunsen flame.

This results in beautiful, different coloured flames – a striking blue-green from copper sulphate, a gorgeous lilac from potassium chloride, a stunning red from strontium chloride, and a comparatively boring yellowy-orange from sodium chloride (you might see this at home if, like me, you are a bit over-enthusiastic sprinkling salt over whatever you're cooking on the stove). The oohs and aahs of delight from my students when they do this experiment are the same ones that are elicited when children see these same colours, produced in much the same way, in a fireworks display.

Flame tests are used to do a form of basic chemical analysis. We can do flame tests on *unknown* substances to find out what they contain – something that burns with a blue-green flame probably contains copper, something that gives a lilac flame is likely to include potassium, and so on. Copper, potassium, strontium and sodium are chemical *elements*, meaning they are substances which cannot be broken down into other substances because they are *made up of only one type of atom*. There are just under a hundred naturally occurring elements and the different colours we see in flame tests are the result of a remarkable property they have: when heated, the atoms of each element give out a uniquely coloured light.

It isn't possible to tell apart the different colours produced by all the elements with the naked eye. Lots of the colours will look the same type of red or orange to us, but if we look at the light from a flame test through a *spectroscope*, which is a device for splitting light up into its constituent colours, we see that the spectrum produced by each element is made up of a distinctive pattern of coloured lines. Each of these coloured lines corresponds to different wavelengths of light, and here's the remarkably useful thing:

> *Each element has its own unique 'fingerprint' of light.*

This was first shown by the inventor of the Bunsen burner, Robert Bunsen, and his colleague Gustav Kirchhoff. It doesn't matter where the light is coming from, whether it's from a fluorescent lightbulb in your kitchen or a firework exploding in the sky, if you see a spectrum with a particular set of wavelengths of light, you can be confident that the light is being produced by something that contains a particular element.

Looking at very detailed spectra of the light from the Sun is what let scientists discover that it contains many of the same elements that we find on Earth. This was a truly ground-breaking discovery – before this, there was no reason to believe that the Sun was made of the same stuff as things on Earth. Looking at the spectra from distant stars confirmed something else for which there had been no proof so far – the Sun and the stars had similar spectra and so were the same type of object.

The great physicist Richard Feynman said that 'the most remarkable discovery in all of astronomy is that the stars are made of atoms of the same kind as those on the Earth', but what's more astonishing is that the link between stars and atoms goes even deeper – the atoms of most of the naturally occurring elements are *made* inside stars, or when stars 'die' in explosions. To understand how this happens, we need to answer another question that the stars compel us to ask – *how do they shine?*

There are few school science experiments where the result is quite as dramatic or memorable as when the two centimetre strip of magnesium you're holding in a Bunsen flame suddenly bursts into a ball of blinding white light. Even seen through the filters that modern health and safety regulations require students to use, this is a spectacular sight. Just for a moment, it's almost as if you've got a miniature Sun at the end of your tongs. If you never did this, you've really missed out – to me, burning magnesium is a rite of passage that every school science student should go through.

I remember being really disappointed that we were given only a single strip of magnesium to share between two of us, and so we had only one go at doing this incredibly exciting thing. When I became a science teacher and found out that these bits of magnesium cost less than a penny each, I decided that I would let my students have two or three pieces each, so that they could do this marvellous, joyous activity more than once.

Burning magnesium is often one of the first *chemical reactions* that students carry out for themselves in school. A chemical reaction is one where two or more substances come into contact with each other and form one or more different substances. When a strip of magnesium is heated in a hot Bunsen flame, it reacts with the oxygen in the air to form magnesium oxide, which is the white ash that is left over once the magnesium stops burning.

Burning, or 'combustion', is called an *exothermic* reaction because it *releases energy* from chemicals in the form of heat. It has arguably been the most useful chemical reaction humans have (so far) discovered and controlled. From burning wood to provide direct heat for warmth and cooking, to burning coal, oil and gas for generating electricity, combustion has been crucial to the development of civilisation.

Most of us know instinctively that the intense white light given out by a piece of burning magnesium is a sign that it is very, very hot. It's not too much of a leap to conclude that stars must also be really hot – which is why, before we knew what stars actually are, some people believed that the Sun was a very big piece of hot metal, perhaps molten iron. *This is typical of how we try to make sense of things we don't understand in the world – we try to explain them in terms of things we do understand.*

However, there's a problem with this: any lump of hot metal eventually cools down, and the Sun shows no sign of this. Another obvious idea is that the Sun must be on fire, that its heat and light come from something that is burning. But again, there is a problem with this idea – even something as big as the Sun would only burn for a few thousand years, no matter what substance it was made of, and we know the Sun is *billions* of years old. The answer to the question of how the Sun shines doesn't lie in chemical reactions, but in a type of reaction we definitely can't do in a school science lab – *nuclear* reactions.

In a chemical reaction, the atoms of two or more substances come into contact with each other, leading to the arrangement of the atoms, how they are joined together, being changed. So, for example, in the case of burning magnesium, we start off with atoms of magnesium in our strip of metal and, in the air, atoms of oxygen joined to each other in pairs as O_2 molecules. Heating the magnesium in a flame provides the energy needed to start the process of combustion, which results in magnesium and oxygen atoms joining together to make magnesium oxide.

There are still the same number of magnesium and oxygen atoms we started off with. The insides of the atoms, their nuclei, have not changed but the arrangement of the electrons around them is different and they are now in a *new* substance where the magnesium and oxygen atoms are attached or 'bonded' to each other.

Substances made of two or more different elements joined together like this are called *compounds*.

As well as combustion, which we can see happening whenever anything is being burned, cooking food, rusting metal, digestion of food, respiration and photosynthesis are just a few of the chemical reactions that are happening around us all the time. But none of these chemical reactions can happen in a star. It's just too hot – the atoms are moving around with so much energy that the forces between them are too weak to hold them together. Instead, what happens inside stars is a result of atoms smashing into each other so hard that their nuclei can join together. Unlike chemical reactions, which start and finish with atoms of the same elements intact, these nuclear reactions result in the formation of different atoms, of different elements.

Nuclear Fission and Fusion

There are two types of nuclear reaction – nuclear *fission*, where a large nucleus splits into two smaller ones, and nuclear *fusion*, where two small nuclei join to make a bigger one. In both types of reaction, the energy released is *millions* of times more than is given out in chemical reactions involving the same amount of 'fuel'. The reason for this lies in the most famous equation in the world, $E = mc^2$, which states that energy (E) and mass (m) are equivalent: energy can be converted into mass and vice versa.

If you completely burned 1 kilogram of coal, it would release about 30 million (30,000,000) joules of energy, roughly equal to the amount of electricity my family and I use at home in one day. When coal is burned, the atoms of carbon of which it is made are still there, but they have combined with oxygen atoms in the air to make carbon dioxide gas. As I hope you might remember from the earlier chapter 'Why Does Ice Cream Melt?', the total mass of the atoms you started with and the total mass of the atoms you ended up with would be the same. If instead you could convert all of the mass of the coal into energy, it would release 90,000,000,000,000,000 (9×10^{16}) joules, enough to meet our electrical energy needs for over 8 million years. In both fission and fusion reactions, there is a decrease in the mass of the nuclei after the reaction has taken place and this mass is given out as energy from the reactions. Although the masses involved in individual nuclear reactions are tiny, I hope you can see from the numbers that even a small amount of mass can be converted into an absolutely huge amount of energy.

Nuclear fission can occur *spontaneously* – a very large nucleus will sometimes split up of its own accord, because it is unstable. However, nuclear power stations, and some types of nuclear bomb, make use of *induced* nuclear fission where unstable nuclei are *made* to split by firing neutrons at them. In a nuclear power station, this process is carefully controlled, so that the energy is released gradually, whereas in a bomb the energy is released in one go.

The fission of 1 kilogram of uranium-235 in a nuclear reactor can be used to generate three million times the amount of electricity produced by burning 1 kilogram of coal, so it is a very useful source of energy. However, just as burning coal results in the production of waste products such as ash and carbon dioxide gas, the fission of nuclear fuels in reactors leaves behind radioactive waste products that have the potential to cause devastating harm to the environment if not stored safely.

I'll come back to the second type of nuclear reaction shortly, but first, I'd like to do with you exactly what I do with my students when I teach this in class: spend a little time discussing the horrific consequences of using nuclear weapons. I do this because I want my students to understand why politicians seem so concerned about other countries doing nuclear tests or developing nuclear facilities. I also want my students to appreciate that science has moral and political dimensions that cannot, and should not, be ignored.

The discovery of the huge amount of energy that could be released by nuclear reactions was immediately recognised for its destructive potential, even before scientists knew how we could deliberately make them happen. In 1904, decades before scientists carried out the first induced fission reactions, the English scientist Frederick Soddy said that if nuclear energy could be 'tapped and controlled, what an agent it would be in shaping the world's destiny!'

At the time, Soddy did not think it was likely that humans would ever be able to control nuclear energy, but that if anyone did, they would 'possess a weapon by which he could destroy the Earth if he chose'. Soddy has been proved correct – the development of nuclear bombs and nuclear power stations have both played a role in shaping the modern world and, sadly, we now have weapons that could effectively destroy the Earth, making it uninhabitable for humans.

I grew up in the 1970s and 80s, a time when it was not uncommon for children to have nightmares that the world would end in a nuclear war. My primary school education and general exposure to news reports were sufficient for me to know that the USA and Russia were the world's super powers and that they were engaged in the Cold War, which meant that they hated each other but were not actually fighting. I also knew that both countries had nuclear weapons and that if the tensions between them mounted sufficiently, one or both countries would use them and start a third, and final, world war.

A nuclear bomb is unlike an ordinary bomb, which relies on chemical explosives, in two key ways. First the explosive power of a nuclear bomb is millions of times more than that of a chemical bomb, so that a *single* modern nuclear bomb could wipe out a city the size of London. Secondly, the neutrons released in the explosion can make material in the vicinity radioactive, and the blast from the bomb scatters radioactive particles throughout the environment, contaminating the air, soil and water, making the surrounding area uninhabitable for years afterwards.

The author Arundhati Roy puts it like this:

If only, if only, nuclear war was just another kind of war. If only it was about the usual things – nations and territories, gods and histories. If only those of us who dread it are just worthless moral cowards who are not prepared to die in defence of our beliefs. If only nuclear war was the kind of war in which countries battle countries and men battle men. But it isn't. If there is a nuclear war, our foes will not be China or America or even each other.

Our foe will be the earth herself. The very elements
– the sky, the air, the land, the wind and water –
will all turn against us. Their wrath will be terrible.

I didn't learn about these things at school, but knew them from reading books and watching films about nuclear war, a common theme in popular culture back then. I particularly remember the television drama *Threads* and the book *When the Wind Blows*, which both vividly depict the consequences of a nuclear war. Works of art like this helped to ensure that the public at the time were kept aware of how horrific and destructive nuclear weapons are.

I don't want my students to live in fear of nuclear war in the same way I did, especially as they are also growing up under the threat of climate change. But I think it's important they should know that, alongside the many benefits of science, it has given us the power to cause devastation and harm in ways that were unimaginable to our ancestors. The question of whether scientists *can* do something often has to be weighed up against the question of whether they *should*, and people continue to debate whether scientists should have helped to develop nuclear bombs.

It is perhaps encouraging that people have only ever used nuclear bombs twice to deliberately kill other humans. Both times, the bomb was dropped by the Americans, on the Japanese cities of Hiroshima and Nagasaki in August 1945, towards the end of the Second World War. These bombs had been developed by scientists working on the US-led 'Manhattan Project'.

One of the scientists working on the project, Robert Oppenheimer, when he saw the first nuclear bomb test, said that he thought of a verse from the Hindu scripture, the *Bhagavad Gita*: 'If the radiance of a thousand Suns were to burst at once into the sky, that would be like the splendour of the mighty one.' Despite Oppenheimer's poetic take, the first nuclear bombs made use of nuclear *fission*, but the process that causes the radiance of Suns is nuclear *fusion*, which has proved to be a much more difficult reaction to replicate and control on Earth.

In a nuclear fusion reaction, two atomic nuclei are forced together to make a new, bigger one. This sounds straightforward enough until you remember that atomic nuclei are positively charged and so repel each other. Trying to bring two protons together is like trying to get a Liverpool fan to sit next to a Manchester United fan – the force required is unimaginable. In order to overcome this repulsion, the nuclei need to be moving incredibly quickly, and for them to be moving fast enough, they need to be at an extremely high temperature and pressure – the kind of temperature and pressure that exists inside stars.

Stars start off their lives as a huge, cold cloud of mainly hydrogen atoms, floating around in space. But you'll remember that the force of gravity acts between all objects with mass, and across any distance, so each atom in this gas is ever so slightly attracted to every other one so that over time, this attraction pulls them together. As they get closer, the size of the force of gravity on them increases, and they 'fall' towards each other, getting faster and faster, just as an object dropped on Earth speeds up as it falls. The faster speed of the atoms means that the temperature of the gas increases as the cloud of gas collapses inwards. Eventually, the gas starts to glow, and a star is born.

Gravity continues to squash the gas together until it becomes a plasma – the electrons are separated from the atoms. After some time, the outward pressure caused by these particles moving around at huge speeds balances the inward force of gravity. The star is in *equilibrium*. Right in the middle of the star, inside its core, is where the plasma is at its hottest. It's here, once the temperature reaches about 15 million degrees or so, that nuclear fusion happens. At these sorts of temperatures, the particles are moving so quickly that they get close enough together for the *attractive* strong nuclear force (which I mentioned in the chapter 'Why Don't Things Fall Up?' and only affects things over a tiny distance) to come into action and overcome the huge repulsive force that like-charged particles have between them.

Under these conditions, hydrogen nuclei are fused together to make a bigger nucleus, and the element helium, named after the Greek word for Sun, is made. All the time this happens, energy is being released, and gradually makes its way out of the surface of the star. It can take thousands of years for energy released in the core of a star to reach its surface. In the case of our Sun, some of this energy makes its way to us, in the form of the waves I discussed in the chapter 'Why Is the Sky Blue?', providing the ultimate source of nearly all the energy we use.

The fusion reactions in a star go on for billions of years – our own Sun is over four and a half billion years old. As well as the energy the Sun provides, its long lifetime is crucial for the development of life on Earth, as we'll see in later chapters. Stars do not shine forever – at some point, the hydrogen in their cores starts to run out, and fewer and fewer fusion reactions happen in the core. Eventually, nuclear reactions start to happen in the layers of the star surrounding the core, causing the star to expand and cool down.

This leads to a new stage in the star's life, where it is known as a *red giant*, because it is much, much bigger than it was. Our own star will go through this phase in a few billion years, and when it does, it will engulf the planets closest to it, including the Earth. In this stage of the Sun's life, the conditions will allow the fusion of hydrogen and helium nuclei to produce bigger elements, such as carbon, nitrogen and oxygen.

What happens next, and how a star 'dies', depends on how much mass the star had when it was 'born'. In the case of a star like our Sun, which has a relatively low mass, the outer layers of the red giant it becomes will blow off into space and gravity will cause the remaining material to collapse inwards to form a *white dwarf* star, which is roughly the size of the Earth. No fusion reactions will take place in this star, but it will continue to give out heat and light because it will be very hot. Eventually, it will cool down and no longer emit much energy, becoming a *black dwarf.*

If a star starts off with a mass about one and a half times or more the mass of our Sun, it will have a very different, arguably more spectacular fate. Such a star will also turn into a red giant, but a much brighter one, with conditions that allow it to produce even bigger elements, sometimes all the way up to iron, which has 26 protons and 30 neutrons in its nucleus. Eventually, this more massive red giant will also collapse, but instead of turning into a white dwarf, it will burst apart in an explosion that is brighter than the light from billions of stars, so bright that it will be seen billions of light years away.

This is a *supernova*. To any astronomers looking in the right direction, it will seem as if a new star has suddenly appeared in the night sky. The temperatures and pressures in this explosion are so high that the big nuclei whizzing about in it can be forced together to make all the naturally occurring elements that are bigger than iron. These elements are blasted out into space by the supernova, in an immense cloud of gas and dust from which new stars, and solar systems, can form. Our own Sun, and all the planets in our solar system, were formed from the remnants of a supernova. I know it's taken a little while for me to get to this point, but I hope you can now appreciate why it's absolutely true to say that we, and everything around us, are literally made of stardust.

As well as producing the heavy elements, a supernova leaves behind other weird and wonderful things:

in the place where the original star of a supernova used to be, the matter left behind will be so dense that a teaspoon of it would have a mass of thousands of billions of kilograms. The gravitational force at this density will force electrons into protons and produce an object that is almost entirely made up of neutrons: a *neutron star*. If the original star started off with more than about three times the mass of our Sun, the neutron star will continue to shrink, becoming denser and denser until it turns into a *black hole* – an object with such a strong gravitational pull that nothing can escape from it, not even the fastest moving thing we know, light.

The physicist Stephen Hawking spent much of his career trying to understand the behaviour of black holes and suggested that they may produce new 'baby universes'. Other scientists have suggested that our own universe is the inside of a black hole. However, unlike stars, where the light they emit allowed us to discover what they are and how they work, we *currently* have no way of knowing what goes on inside a black hole.

The Big Bang Theory

The scientific answers to the questions 'What are stars?' and 'How do they shine?' would have delighted those early astronomers who suspected we could never really know, but none of them could have imagined that studying the stars would provide us with an answer to perhaps the biggest question of them all: *where did the universe come from?*

Humans have pondered this question throughout our history, and pretty much every culture that has existed has come up with an answer of its own. We refer to these answers as creation *myths*, but the truth of these stories was not always questioned and, even today, there are some people who believe that the creation story of their particular religion is literally true. One thing a lot of these myths have in common is the idea of a *creator*, a *supernatural* being who *makes* the universe, often by bringing 'order' out of 'chaos'.

The story science tells us about how the universe began is known as the *Big Bang Theory* and it does not involve a creator. There are other fundamental ways in which it is different and, as I will try to explain, it is by understanding these differences that we can argue that it is the *best* explanation we have for the origin of the universe.

The Big Bang Theory says that the all the matter *and space* in the universe was once condensed in the same point, existing as an inconceivably small and dense object which exploded about 14 billion years ago. The stuff that was released from this explosion eventually turned into stars, galaxies, planets, *us* and everything else that exists, through a series of *natural, physical processes.* This may seem rather dull and lacking in the characters, plot and drama of other creation stories, but coming up with this version of how the universe started arguably took far more imagination, creativity and effort than all of the other creation stories combined.

The Big Bang Theory was partly developed by a Belgian priest, Georges Lemaître, whose contributions to science are amongst those of many who have demonstrated that religious beliefs are no barrier to excellent scientific thinking. Lemaître had carried out some calculations using Einstein's theory of general relativity and come to the conclusion that, if Einstein's theory was correct, the universe must be expanding. He then argued that if the universe was expanding today, then it must have been smaller back in the past.

Taking this idea to its logical conclusion, the universe must have started off as a very small, very dense 'primeval atom'. Einstein rejected this idea when he first heard about it, calling it 'abominable' – he believed that the universe was 'static', that it had always existed just as it was. But only six years later, in 1933, Einstein was overflowing with praise for Lemaître, saying that his idea was 'the most beautiful and satisfactory explanation of creation to which I have ever listened'. So, what changed Einstein's mind? Evidence.

The discovery that started the process of scientists, and then wider society, accepting the Big Bang Theory was made by the astronomer Edwin Hubble at the end of the 1920s. Hubble used spectroscopy, the same method Bunsen and Kirchhoff had used to study the composition of stars, to study the movement of galaxies. He found that they were moving away from us, and that the further away from us they were, the faster they were moving. This supported Lemaître's idea that the universe was expanding, and that everything in it had started off much closer together.

The Big Bang Theory tells us that galaxies are not only moving away from us through space, but that the space between galaxies is expanding. This may be hard to get your head around, but you can get the idea from this simple demonstration:

Take a rubber balloon and stick some paper dots on it before you blow it up. As you blow the balloon up, the dots move farther away from each other, not because they are moving across the surface of the balloon, but because the surface of the balloon is expanding, increasing the distance between the dots.

Another crucial feature of the Big Bang that can be seen by doing this balloon demonstration is that, if you pick any dot to look at, you'll notice that all the dots move away from it – there is no centre on the surface of the balloon from which the dots are moving away and similarly, there is no central galaxy or star in the universe.

Hubble's discovery about the moving galaxies was not sufficient to convince everyone that the universe was expanding. Further evidence to support Lemaître's idea came from the fact that it *correctly predicted* the amount of hydrogen and helium in the universe – something that scientists had already confidently calculated. To become a fully-fledged scientific theory, Lemaître's idea needed to do one more thing: predict something about the universe *which had not yet been observed.*

In 1948, Ralph Asher and Robert Herman predicted that if the universe had started as Lemaître suggested, then there ought to be some electromagnetic radiation left over from the big bang, and that this would be in the form of microwaves with a wavelength of about 1mm. They argued that the radiation should be detectable no matter which direction you looked out into space, because it was coming from every possible direction. This 'Cosmic Microwave Background Radiation' (CMBR) was detected in 1965 by Arno Penzias and Robert Wilson, and quickly led to the Big Bang Theory being accepted by most scientists.

The details of precisely how the Big Bang Theory was developed, and the mathematical models that go along with it, are complicated, but I think the basic idea is as straightforward to understand as any creation myth. A better name for it would be the Big Bang Model because, as with the theory of evolution by natural selection (which is covered in the later chapter, 'Are Fish Animals?'), some people think that the word 'theory' means these ideas have somehow *not* been proven by science in the same way as, say, Newton's *laws*, or the *facts* about things that have been discovered through science.

This is not entirely the fault of these people – 'theory' has a very specific meaning in science, which is almost contradictory to its use in everyday conversation, where it can be used to mean a guess or a belief. You can have a theory that Liverpool would win the Premier League if they changed their attacking line-up, or you can believe in the conspiracy theory that vaccinations are a way for governments to have us all implanted with microchips. Neither of these uses of the word 'theory' corresponds to its application in science.

A *scientific* theory is a set of ideas that provide a coherent, thorough explanation for things we see in nature, provide evidence for that explanation, and, perhaps most importantly, provide a way of making predictions about the aspect of nature it explains *which can be tested.* These are all things that are missing from creation myths, but that the Big Bang Theory does beautifully.

The stars inspire a sense of awe and wonder in most humans, and it is perhaps this which has driven so many people to devote their working lives to understanding these once mysterious objects. However, for me, the real wonder lies in the fact that the atoms that were forged in the heart of a star and in a supernova can come together to form beings that are capable of asking, and indeed answering, the question, 'What are stars?'

Chapter 6
Are Fish Animals?

I had a friend at primary school whom teachers and other adults described as 'wild'. His parents weren't around much, so at weekends and during school holidays he was usually the first one out to play, and the last one who went home. He had a carefreeness about him that was magnetic to other children, and he was often the instigator of the mischief we got up to when there was nothing else to do. If you'd met this boy, you would never have guessed that, hidden away under his bed, he had something which brought him a sort of joy that couldn't be found in playing football or climbing trees: a stamp collection.

Inside a special album, with pages marked out in grids, he had painstakingly organised hundreds of stamps according to their country of origin, their value and other categories which he had decided upon. While this hobby may have seemed surprising for someone like him, my friend wasn't unusual in that many children go through a phase of collecting objects, often toys or picture cards of some sort. Even without any help or encouragement from adults, children will also *organise* their collections, maybe in order of size, or type of toy, or using some other system of *classification* only they can explain.

Imposing some sort of order on their collections seems to be part of what makes collecting things fun and satisfying for children, and indeed adults. This human urge to arrange things into groups, and put them into categories according to some kind of specific criteria, is arguably one of the key driving forces of *science* – much of our scientific understanding of the world relies on collecting information and organising it.

The practice of classifying things is known as *taxonomy*. In school science, children are taught how to classify everything from stars to rocks to molecules, based on the differences between things like main sequence and white dwarf stars, igneous and metamorphic rocks, and organic and non-organic molecules. But it is in biology, the science of living things, where classification takes centre stage, because the way we classify living things, or 'organisms', is central to our understanding of them.

Young children (and even many adults) may ask a question like 'Are fish animals?' because, regardless of whether they've had any biology lessons or not, they learn early on that living things come in a huge variety, and that they are classified into groups such as animals, birds, fish, insects, mammals, and so on. The confusion children may have about whether fish are animals is totally understandable. Fish live in water and they don't have limbs – just two of the striking ways in which they are significantly different from creatures such as cats and dogs, which children soon become familiar with as 'animals'. This illustrates something that has been a problem for scientists: which features should be taken into account when classifying living things into different categories?

The precise way in which scientists today categorise organisms into groups is not obvious. How scientists decide whether something is an 'animal' or not is based on a set of rules which people have *invented* and, like the rules of a game we don't know, we need to be *taught* them before we can 'play' properly. I'll discuss the system we use to classify living things later in this chapter, but first we need to start with a classification problem that may seem easy to solve at first, but which can prove quite difficult once you start thinking about it: what is the difference between a living thing and a non-living thing? Or to put it another way, 'What is life?'

This is a fundamental question in science and one that we need to answer precisely if we want to confirm the existence of life somewhere other than on Earth, or if we ever have to decide if a computer or other human-made thing should be treated as if it is 'alive'.

One way to start answering this question is to think about which characteristics living things have that non-living things don't. If you try to do this yourself, as many schoolchildren are encouraged to do in their early biology lessons, it quickly becomes apparent that it's difficult to find a *single* property which *only* living things have. This illustrates one of the key challenges in biology – things are rarely as clear-cut, or as neatly defined, as we might like scientific definitions or classifications to be. When it comes to defining 'life', or what makes something 'living', it may surprise you to know that scientists themselves are still not entirely agreed upon a straightforward answer.

There's no simple way to differentiate between a living thing and a non-living one. Instead, schoolchildren are taught a list of seven *processes* which, taken

together, allow us to identify whether something ought
to be classified as a life form:

- *Movement* – the ability to move (change position)
  by themselves.
- *Respiration* – the chemical processes by which living things obtain energy from food (note that this
  is *not* the same as breathing, which is the process
  by which some animals get air in and out of their
  lungs).
- *Sensitivity* – the ability to detect and respond to
  changes in their environment.
- *Growth* – the ability to get bigger or develop in
  other ways with age.
- *Reproduction* – the ability to make more living
  things of the same type.
- *Excretion* – the production and getting rid of waste
  products.
- *Nutrition* – the need to take in and make use of
  food.

These processes are usually listed in this order so that
their first letters spell out MRS GREN (some teachers use MRS NERG). It's one of several mnemonics
that crop up in a typical science education to help
students remember everything from the order of

the colours in the visible spectrum (Roy G. Biv or Richard Of York Gave Battle In Vain) to the order of the planets from the Sun (My Very Easy Method Just Speeds Up Naming).

Sometimes, MRS GREN is taught as MRS C GREN or MRS H GREN, with the C standing for 'control' and the H for 'homeostasis', the process by which living things control the conditions inside their bodies to keep them stable. This is why your body can automatically adjust its internal temperature by sweating when you're too hot, or shivering when you're too cold.

The MRS GREN characteristics are useful for getting us to think about what makes something living or non-living (as opposed to 'dead', which is something that only once-living things can be), but they're not an absolutely definitive way of identifying whether something is, or was, 'alive' in the scientific sense. As you may have noticed, fire could meet the criteria for a living thing as defined by MRS GREN – it moves, it needs 'food' (fuel) and oxygen, it grows, it 'reproduces', and so on.

Someone who knows a lot about animals might point out that mules, which are very clearly alive, cannot reproduce (mules are usually born as a result of a male donkey mating with a female horse; when two mules have sexual intercourse, it does not usually lead to pregnancy). So, after having covered MRS GREN in school science, children should eventually be taught a more accurate way of determining whether something is a form of life as we know it:

All living things (organisms) are made up of one or more cells.

I've already described in 'What Is the Smallest Thing?' how schoolchildren are given the opportunity to see cells using microscopes, and how the invention of this instrument was crucial in the development of our scientific understanding of the world, perhaps no more so than in biology. We can think of cells as being the 'atoms' of biology, they are the basic unit, the fundamental building block, of life. Cells are the smallest things that have all the characteristics of living things.

All the processes of life described above are ultimately the results of what happens inside cells; for example, organisms grow when their cells split and re-expand, and respiration is a chemical reaction that takes place inside cells. This is known as the *cell theory* of life and it is as central to understanding living things as Newton's theories, and the particle model of matter, are to understanding the physical world.

I'll be coming back to cells in more detail in the chapter 'What Am I Made of?' But first, let's get back to MRS GREN and another problem with it that I have encountered as a teacher: it's not at all unusual for children to learn about these characteristics and soon forget, or fail to have understood in the first place, that *plants* meet all these criteria and are living things. I've met more than one teenager who has been surprised when I've talked about plants as being 'alive' and I know other teachers who have experienced this too. It's not hard to see why this might be the case if we think about MRS GREN and how these processes are mostly *not visible* in plants.

For example, plants tend to be rooted to one spot, and it's not obvious that they can move on their own or that they respond to changes in their environment. To convince children that plants can indeed do these things, many school biology departments have a Venus flytrap or Mimosa, plants that have leaves which move quite dramatically when touched. Another way to show children (or convince yourself) that plants move is to watch a time-lapse film of daisies, which move slowly through the day to follow the position of the sun in the sky, then close their petals at night.

We can't see *respiration* happening in animals or plants, but we can see animals breathe to get the oxygen they need for this MRS GREN process, something plants do not do. Similarly, we can see animals eating to provide themselves with *nutrition*, and excreting the waste products but again, eating and pooing is not something plants do.

Photosynthesis

Contrary to what the Ancient Greeks believed (and some people still believe today), plants do not get much of their food from the soil either. This was first shown in the 1600s by the Belgian scientist Jan Baptista van Helmont, who planted a small willow tree in a pot with about 90 kilograms of soil and showed that as the tree grew and increased in mass by over 70 kilograms, the mass of the soil barely changed at all.

This was the first of many experiments which would lead scientists to eventually understand that the main source of a plant's nutrition is a process called *photosynthesis*, a chemical reaction that combines water (usually taken up through roots) and carbon dioxide from the air (absorbed through the leaves) to make a sugar called 'glucose'. The reaction can be summarised in this simple 'word equation', which students are often asked to memorise:

carbon dioxide + water → glucose + oxygen

You can see from the equation that, as well as glucose, photosynthesis produces oxygen, which is released as a gas. This is usually shown to students in one of the few experiments in school biology that doesn't require cutting things up or waiting ages for something to happen. Instead, it involves collecting a gas as it bubbles up from a green plant (usually some form of pondweed) placed under water. You can prove the gas given off by the plant is oxygen using a test taught in chemistry lessons:

> *Take a burning splint (thin bit of wood) and blow out the flame, then put the still-glowing end of the splint into a test-tube of the gas. If it's oxygen, the splint will re-light and you'll once again have a flame.*

Putting the plant closer to a source of light (usually a simple desk lamp) speeds up the production of bubbles and provides a convincing demonstration that light intensity affects the rate of photosynthesis.

As with so many school science practicals, this one is based on experiments done by scientists hundreds of years ago. In this case, similar experiments were carried out in the 1700s by the English scientist Joseph Priestley, who showed that plants produce oxygen, and the Dutch-born Jan Ingenhousz, who was the first to show that photosynthesis needs light to take place. The word 'photosynthesis' comes from the Greek words for 'light' and 'putting together', but there is another crucial ingredient plants need before they can make their own food: chlorophyll.

This is the chemical inside plant cells that gives them their green colour, and allows energy from light to be absorbed and used for photosynthesis. Sometimes this is shown in the above equation by writing the word 'light' above the arrow and 'chlorophyll' below it, because they are necessary for the reaction to take place.

It is no exaggeration to say that photosynthesis is perhaps the most important chemical reaction on Earth: it is the ultimate source for nearly all the food and oxygen that living things on our planet need to survive. At a time when humans are desperately trying to find better ways to capture energy from the sun and *store* it, using things such as solar panels and batteries, it's astonishing to think that nature had a mechanism for doing this over a billion years ago.

Photosynthesis literally allows plants to store 'solar energy' in their cells. This energy is then passed onto animals that eat plants, and then to the animals which eat those animals, and so on. The fossil fuels (coal, oil and gas) from which we currently extract most of the energy we need to run our civilisation, are the remains of living things, so the energy stored in them can also be traced back to photosynthesis. The oxygen released in photosynthesis is just as important – it allowed the development of the complex lifeforms we have on Earth, as well as the formation of the ozone layer, which protects us from harmful radiation from the sun.

The ability to photosynthesise, to make their own food, is the most significant thing that sets plants apart from other living things. But even if we didn't know about this, there are plenty of other differences between plants and animals that make dividing living things into these two categories an obvious thing to do. 'Obvious' is the key word here – it means 'easily seen' – and early attempts at biological classification relied entirely on *looking* at living things and sorting them according to what their bodies looked like, inside as well as outside.

These days, dissection, the cutting up of a once-living thing to study its insides, is not a required part of most school science courses. Many young students would find it distressing to dissect an animal. Others would object on moral grounds. My own view is that there is no real need to cut up a dead frog or other animal when we have access to realistic models and computer simulations.

However, I think it's important to know that dissection played a significant role in the development of our biological knowledge. I also think that students should experience it for themselves by cutting up something like a flower, because it helps reinforce that science is a *practical* activity and that the first step towards understanding, or discovering something new, is often *observation*, the act of looking very closely.

The invention of the microscope gave scientists a way of looking even more closely at living things, and another way to differentiate between plants and animals – by examining their cells. Plant cells have a distinct 'wall' around them, which animal cells do not. Sea anemones look a lot like underwater flowers and they tend to stay anchored to one spot, but their cell structure means we classify them as animals. Today, as well as using straightforward observations to classify newly discovered organisms, scientists can use DNA (more about that in the next chapter) to make more precise classifications. This is important because there are literally millions of different kinds of living things on our planet and it's useful to have more than the two categories of 'plant' and 'animal' to put them in.

Long before there was anyone we might describe as a 'scientist', humans had subdivided plants and animals into smaller categories of organisms with shared characteristics, such as trees, herbs, birds, fish and insects. Each of these categories can be further divided into smaller ones, based on observable differences, such as evergreen and deciduous trees or flying and flightless birds. If you carried on with this process, you would eventually end up with organisms that are similar to each other in almost every possible way – a *species*. One way in which species are defined is as a group of organisms that can reproduce with each other. Again, as with a lot of categories in biology, there are exceptions called 'hybrid species' such as 'ligers', which are produced when male lions mate with female tigers.

It might take you a while but you could, if you wanted, come up with your own way of classifying living things, starting with individual species and grouping them together in bigger and bigger categories, based on shared features like having hair (all mammals) or feathers (all birds), or whether they have a backbone (all vertebrates), until you got back to plants and animals.

Whilst it might be fun or interesting to do this, if we want to share knowledge and information about living things with other people, if we want to do *science*, we need a universally agreed-upon classification system. The one scientists use today is largely based on one devised in the 1700s by the Swedish botanist and zoologist Carl Linnaeus and known as *Linnaean Taxonomy*.

Linnaean Taxonomy

Linnaeus is sometimes referred to as 'the father of taxonomy' for his work. He was also known as Carl von Linné later in his life, a title he was given by the King of Sweden, who made him a lord as a reward for his accomplishments. Historians usually describe Linnaeus as a botanist and zoologist, because this is how he would have been described in his own lifetime, before the word 'scientist' was invented in 1834.

Despite the fact that he may not have known the word himself, Linnaeus embodies one of the qualities that I think is common to all people who become scientists or who love science for the sake of it: a limitless fascination with the natural world. I think it's maybe something we all have the capacity for, but which can quickly be ground out of us if it is not nurtured or encouraged. It seems obvious to me that luck and privilege play a role in the extent to which children are given opportunities to engage with science, just as these things can determine whether someone grows up with an appreciation of art, or music or literature.

In Linnaeus's case, like many of the early scientists whose work laid the foundations of modern science, he was from a family background that allowed him to indulge his interests, to get an appropriate education and have other opportunities which led him to develop and share his ideas about nature. We often remember these people as 'geniuses', but I have no doubt there were lots of other people that have lived throughout history who could have accomplished the same things if they'd had the same opportunities.

Linnaeus collected plants, animals and minerals and first published his own classification system in 1735 in a short book called *Systema Naturae* (Latin for 'System of Nature'). The ideas he presented in it were quickly recognised as being useful, and were adopted by other people doing similar work. Over the next thirty-five years or so, he continued to collect plants, minerals and animals, and updated the book and his ideas more than ten times. Since then, scientists have changed the details of how they classify living things as they have learned more about them.

For example, since Linnaeus's time, scientists have agreed that fungi, such as mushrooms, ought not to be classified as plants because, despite usually being rooted to one place and their cells having walls, they do not photosynthesise, and what happens inside their cells is more akin to what happens inside animal cells. Similarly, as scientists have learned more about single-celled organisms (living things made up of only one cell), they've seen that it makes no sense to divide these into the two categories of 'plant' or 'animal', and have given them categories of their own.

Whether you remember learning about Linnaean taxonomy at school or not, you probably know that the scientific name for humans is *Homo sapiens*. This way of naming organisms is called the *binomial system* (because it uses two names) and Linnaeus is credited with establishing it as the system scientists everywhere use for naming all organisms. I explained in the chapter 'Why Does Ice Cream Melt?' how the introduction of a systematic way of naming chemicals, a *nomenclature*, helped improve the science of chemistry, and the same is true of biology. Having a standardised way of naming animals means that every species has a unique scientific name by which it can be identified.

This is an obvious improvement over having different names for organisms in different languages or even in the same language in different places. But this is not the only benefit of naming organisms in a systematic way – just as chemical nomenclature allows us to know some of the properties of a chemical from its name, binomial nomenclature tells you something about the organism.

The first part of an organism's binomial name tells you which *genus* the organism belongs to and the second part tells you the *species*. These are the two 'bottom' categories in the Linnaean system of classification, which is a *hierarchical* way of categorising organisms, meaning that the groups are put in a ranked order of decreasing size in terms of numbers of organisms. A hierarchical way of organising things is common in lots of places – in a well-stocked supermarket, a particular type of burger bun (say, my favourite, a sesame-seed covered brioche bun) should be found on a section of a shelf with other buns in the bakery aisle with other baked products, which itself should be in the food section of the supermarket.

The equivalent of the supermarket in the Linnean system is 'all living things', which are first split into five *kingdoms*:

- animals
- plants
- fungi (mushrooms, yeasts and moulds)
- protists (single-celled organisms with a cell nucleus)
- prokaryotes (single-celled organisms without a cell nucleus)

Each kingdom is then split into six further divisions:

- phylum
- class
- order
- family
- genus
- species

Since I've already mentioned a few mnemonics, I should probably tell you that there are lots of them to remember this hierarchy, including King Philip Came Over For Good Soup. Each of these categories, from kingdom to species, has sub-divisions of its own; for example, one of the sub-divisions of phylum is *chordata* (mainly animals that have backbones), which can then be divided into a *class* of mammals, birds, amphibians, reptiles or fish.

I'm not going to list every subcategory as this isn't a textbook, but I think it is worth looking at examples of how a couple of animals are classified. First, orangutans:

Kingdom: Animal
Phylum: Chordata (vertebrate)
Class: Mammal
Order: Primate
Family: Hominid (great ape)
Genus: Pongo
Species: pygmaeus

Now humans:

Kingdom: Animal
Phylum: Chordata
Class: Mammal
Order: Primate
Family: Hominid (great ape)
Genus: Homo
Species: sapiens

One of the benefits of classifying living things in this way is that it summarises information about them – the fact that an orangutan is a mammal, for example, would tell a scientist who knew nothing about orangutans that they have backbones and hair or fur, and that they have mammary glands which, usually in females, can produce milk to feed new-borns. This sort of classification can also reveal how similar organisms are to each other. As you can see from the way they are classified above, humans and orangutans have enough characteristics in common to belong to the same *family* of organisms.

This might not seem like a big deal to you, but in the 1700s it was controversial to suggest that humans could be put in the same category as animals, or that they could be related, because of the widely held belief that humans were created by a god in his image. However, less than a century after Linnaeus died, scientists would begin to see how classifying living things in this way, based primarily on observations about how they look, hints at the true nature of life on Earth and how *all* organisms are related to each other on a fundamental level.

Unfortunately, there have been some unpleasant consequences of classifying living things, and it would be remiss of me to ignore them before proceeding.

As well as putting animals into groups and hierarchies, Linnaeus did the same with humans. It saddens me that we live in a world where I probably don't have to tell you which people he put at the top of his hierarchy and which at the bottom. According to the *Linnean Society of London*, which keeps original copies of much of Linnaeus's work, 'Africanus consistently remained at the bottom of the list. Moreover, in all editions, Linnaeus's description of Africanus was the longest, most detailed and physical, and also the most negative.'

Linnaeus described entire populations of people using negative stereotypes based on little more than his personal ignorance and prejudices. But the real harm came from the fact that his reputation and status gave weight to these views, and helped spread them. Our modern understanding of human biology proved Linnaeus wrong a long time ago. It is abundantly clear that all humans are members of a single species, and that there is no biological basis for 'race'. In fact, science tells us that two people in Africa can be more different at a genetic level than an African and a European. Despite this, people continue to try to support their *racist* views of other humans using science.

For example, there are people involved in education who believe that children from certain ethnic backgrounds underachieve in school because of 'genetics'. These people don't always have a background or training in science, but that's kind of my point: science has an *authority* that even non-scientists want to borrow in order to add credibility to their beliefs. History is littered with examples of wars and genocides caused by people believing other people are *literally* inferior to themselves.

Whilst we might want to celebrate Linnaeus's contributions to science, it would be wrong to fail to acknowledge that scientists like him have played a role in validating deeply harmful notions about the differences between people, and perpetuating racist and sexist ideas. Categorising living things may be useful to scientists, but the disadvantages some people face are often due to the problematic ways in which we categorise humans from different backgrounds.

The Theory of Evolution

This chapter started with a question to which we now have an answer – fish are indeed animals, at least according to the scientific system of classification we use to put living things into categories. At this point, I worry that what I've written so far may have reinforced some negative stereotypes about school biology being just about remembering the names of things and being able to label diagrams correctly. It's sad that the subject has this reputation and, by the end of this chapter, I hope to convince you that it is undeserved.

It's true that biology as a science developed because early 'naturalists' and 'botanists' like Carl Linnaeus did a lot of work simply finding different organisms and giving them names. But, as I've said before, this is really the first step in science – taking an interest in something and looking really closely. Finding and cataloguing different organisms were the *necessary* first steps that led to scientists asking, and eventually answering, a question with far-reaching implications for our understanding of the world and our place in it: why are there so many different kinds of living things and where did they all come from? Or, to put it in the words of Charles Darwin, one of the scientists who would answer the question, what is the *origin of species*?

Darwin's answer to this question, his *theory of evolution by natural selection*, has been described as 'the best idea anyone has ever had'. It's certainly a top contender for the most *famous* scientific idea in the world. It's also perhaps the scientific idea that is most widely misunderstood and widely *rejected* by people. I've already mentioned the first misconception about it in 'What Are Stars?' – the fact that it is called the *theory* of evolution by natural selection. The use of the word 'theory' leads a lot of people to believe that it is *just an idea*, some kind of *guess* at answering the question of how life on Earth developed.

I've explained this previously but, as I do when I teach these subjects in school, I think it's worth reminding you that the word 'theory' in science does *not* have the same meaning as in its everyday use. A *scientific* theory is a set of ideas that provide a coherent, thorough explanation for things we see in nature, provide evidence for that explanation and, perhaps most importantly, provide a way of making predictions about the aspect of nature it explains, *which can be tested*. Just as Newton's theory of universal gravitation explains how the force of gravity works, Darwin's theory of evolution by natural selection explains how biological evolution works.

I've added the word 'biological' in front of evolution because I want to avoid spreading another common misconception about this subject – the idea many people have that evolution is related to something improving, getting better. The manufacturers of mobile phones and other gadgets might claim that their products have 'evolved', and I hope you might think that reading this book has helped your understanding of science to evolve. In this context, the word 'evolve' is synonymous with 'progress' and this is what many people, through no real fault of their own, think it means in a scientific context too.

The iconic image of evolution is a picture showing five or six figures in a line, starting with some sort of ape crouching on all fours on the far left and ending with a modern human standing tall on the right. It is a clear depiction of positive development in a straight line from one figure to the next. The image is commonly known as known as the 'march of progress' and is based on an original book illustration called 'The Road to Homo Sapiens'.

There are countless versions of it on everything from the pages of textbooks to the front of T-shirts, but I'm pleased *not* to include it in this book because, whilst it has helped embed the idea of evolution into popular culture, this image has also, without doubt, helped spread the incorrect idea that the evolution of life follows a straight path towards us humans as the very pinnacle of some kind of grand plan. This may be an appealing idea, but it is not true.

In biology, evolution refers simply to the fact that living things *change* over time, or to put it more precisely:

> *Evolution is a change in the inherited characteristics of a species over successive generations.*

This idea that living things evolved was around long before Charles Darwin. The fact that he came up with the best theory to explain it was no doubt influenced by the work of his grandfather Erasmus Darwin who, before Charles was even born, wrote down his own thoughts on evolution in his book *Zoonomia*. Erasmus's ideas, published in 1794, were controversial because they contradicted the widely held religious belief that God had created all living things in exactly the state that we see them today. If he was right, the Bible would have to be wrong, and this was an unacceptable idea at the time.

However, around the same time, enough living and once-living things, such as fossils, were being catalogued for more people to be convinced that organisms changed over time. Other scientists, including the French zoologist Jean-Baptiste Lamarck, started to put forward ideas about how it happened. In a book published in 1809, Lamarck suggested that animals changed within their lifetimes, perhaps to cope better with their environments, and that these changes were passed on to their offspring.

The classic example of 'Lamarckian evolution', which children still learn about in schools today, is of ancient giraffes repeatedly stretching their necks to reach leaves higher up in a tree, thus elongating their necks and going on to have offspring born with similar longer necks. You can perhaps see a way to test this – change some physical characteristic of an organism and see if the change is passed on to its children.

This is exactly what the German biologist August Weismann did, by cutting the tails off several generations of mice to see if there was a decrease in the average length of their tails from one generation to the next. You probably won't be surprised to learn there was not any change in the average tail lengths of mice descended from the mice who'd had their tails chopped off – Weismann had *disproved* Lamarck's theory, something no one has yet managed to do with Darwin's.

Natural Selection

Today, people who don't accept that life evolves are in a minority, and most scientists accept that the *process* by which biological evolution occurs is *natural selection.* This is an idea that was first put forward by Charles Darwin and Alfred Russel Wallace in 1858, in a paper called *On the Tendency of Species to form Varieties; and on the Perpetuation of Varieties and Species by Natural Means of Selection* in the *Zoological Journal of the Linnaean Society.*

Both scientists had separately come to the conclusion that, just as farmers and animal breeders could cause changes in plants and animals by selective breeding (or *artificial* selection), a *natural* process of selection could be responsible for the changes observed in living things over time.

It's important to note that there is no sense in which natural selection *designs* living things to have particular characteristics. The theory of evolution by natural selection does *not* require this process to be *motivated* in the same way a dog breeder might have a motivation to create a breed of dogs with longer ears or shorter legs (both of which have been done). Instead, what happens is something like this:

- In any environment, for any particular species, there are *limited* resources, such as food and other members of the opposite sex to mate with. So, there will always be competition for resources and a 'struggle for survival'. This is why populations of living things cannot keep growing indefinitely.

- Individuals within any population (group) of a species are *varied* and have different characteristics or 'traits'.

- Because of these differences, some individuals in a population will have advantages over others in terms of being better-suited to their environment, so they will be *more likely* to live longer, breed and pass on their genes (including the ones that cause the advantageous traits).

- Effectively, these traits have been *naturally* 'selected' to be passed on, so become increasingly common.

- Over many, many generations, the repeated process of natural selection, acting on a variety of traits, leads to organisms becoming so different that they can no longer reproduce with organisms like the ones from which they are descended – they are now a new species.

If you're not familiar with the word 'genes' above, don't worry for now – all you need to know is that, just as the idea of particles as 'units of matter' helps us to understand the behaviour of physical substances, the idea of genes as 'units of heredity' helps us understand how traits are passed on from parents to children. I'll explain a bit more about them in the next chapter.

Darwin and Wallace were unaware of the existence of genes, so they didn't use quite the same words as I have above, but they did come up with the basic process that explains how evolution occurs and how a new species can be formed from an existing one. The process above is often described as 'survival of the fittest' which, unfortunately, is another term related to this theory that lends itself to misinterpretation. In the context of evolution by natural selection, 'fittest' simply means 'best adapted to the environment', not necessarily the strongest or fastest, or any of the other qualities usually associated with being 'fit'.

Darwin also used the phrase 'descent with modification' to describe evolution and the fact that offspring (of all organisms) were descended from parents but were not exactly the same as their parents; they were 'modified'. Today, we know that the way in which children inherit characteristics, like eye colour and blood type in humans, is through the transmission of *genes*. At a fundamental level, genes are long strings of chemicals with a particular arrangement of atoms that encode *information*. It is the *information* which is passed on from parents to offspring that determines which characteristics the offspring will have.

However, just as information in real life can get muddled up as it is passed on from one source to another, the information in genes can occasionally, *randomly*, be *mutated*. When this happens, it can either have no overall effect on the organism they belong to, or result in an advantage or disadvantage in terms of the organism's ability to survive and reproduce. Over billions and billions of years, these tiny, random mutations in genes are what eventually lead to lifeforms changing, and ultimately to the wide diversity of life we see on Earth.

Although Darwin and Wallace both came up with the idea, the reason why Darwin is the name most of us associate with evolution by natural selection is that he published a book in 1859, *On the Origin of Species*, in which he presented a more detailed argument with lots more evidence. The book became a bestseller and took these ideas to a much bigger audience of the general public, not just scientists.

It's important to acknowledge Wallace's contribution, not only because he fully deserves the credit, but because co-discovery is not unusual in science when lots of people are thinking about the same problem at the same time. More importantly, the fact that two people had independently come to the same idea suggested that the process of natural selection was a real phenomenon that was 'discovered' rather than invented.

One conclusion we can draw from the theory of evolution by natural selection, and it is one that is now accepted by the majority of scientists, is that *all* life on Earth is descended from a *single* original life form. This organism, from which all other organisms evolved, probably lived about 4 billion years ago and is given the name *Luca*, an acronym for 'Last Universal Common Ancestor'. It is not easy to believe in the existence of Luca, not least because it is almost impossible for us to imagine the slow, cumulative changes that took place over billions and billions of years to get from a single-celled organism to a human, or even, say, a beetle.

Our planet, Earth, is about 4.6 billion years old. That is 4600 million years. The first simple, single-celled lifeforms appeared 4 billion years ago. It took another billion years before photosynthesising cells emerged. Multicellular life, made up of more than one cell, has only been around for about a billion years, and simple animals for just over half of that. Mammals appeared about 200 million years ago, and anything we might call a human less than 10 million years ago. If this whole time was to be condensed into only 24 hours, humans have existed for less than two minutes.

As well as the mind-boggling lengths of time involved, another intellectual hurdle for many of us in accepting evolution by natural selection is that, to our ordinary way of making sense of the world, it can *appear* that living things are *designed* to be the way they are. It is intuitive to think that something as complex as an animal *must* have been designed, that there *must* be some sort of mind or 'creator' behind it all. In fact, to people throughout history, and to many people today, the apparent *order* and complexity we see in the world is proof of a god who designed things to be the way they are.

In theology, the study of religion, this is known as the 'teleological argument' or 'argument from design' for the existence of a god. One of the most famous and convincing versions of this is the 'watchmaker analogy' devised by the Christian preacher William Paley. In his book *Natural Theology – or Evidence and Attributes of the Deity Collected from the Appearances of Nature*, Paley argued that if you had never seen a mechanical watch but came across one by chance, you would look at its complicated construction and conclude that it must have been designed and made by someone who had a purpose for it. He then goes on say that nature is full of examples, such as the human eye, which show exactly this sort of design.

It is a powerful argument, and one that is difficult to refute unless you are equipped with much more information than anyone had at the time Paley wrote his book. Although it was published decades before Darwin's work, Paley's analogy was used widely, and continues to be used even today, by people wanting to discredit the theory of evolution by natural selection (I've even seen a version with a table-tennis ball instead of a watch presented on a TikTok video). Darwin argued that Paley was wrong, and explained that something like the human eye could have developed from much simpler ones through the process of evolution by natural selection, and there must have existed in nature *intermediate* versions of the human eye that were progressively more useful to the organisms which had them.

Intermediate Forms

Today, we know Darwin was right, not least because the intermediate forms of the eye he predicted to exist have been found in various living and once-living organisms. There are hydra (tiny animals which live in water) that have only a few photosensitive cells on their bodies, which allow them to detect the presence of light. There are also some jellyfish that have similar photosensitive cells in dents, or holes, in their bodies, which has the advantage of letting the jellyfish know what *direction* the light is coming from. Likewise, scientists have discovered there are other organisms, such as snails, which have (or had) eyes that can be considered 'intermediate' forms of the kind of eyes we have.

Octopuses arguably have *better* eyes than us, because they don't have a blind spot. If you're in any doubt that an intermediate eye would be of any use, you could, as I've already suggested in the chapter 'Why Is the Sky Blue?', make yourself an intermediate or pinhole camera (with no lens) and look at the image you get when all you have is a dark box with a hole at one end and a screen at the other. You won't get images as clear or detailed as you would with a proper camera, but you might be surprised by how good they can be.

The existence of intermediate forms is one of the key pieces of evidence that supports the theory of evolution by natural selection. We can see these in living creatures, and in fossils, which are the preserved remains of ancient organisms, usually found in rocks.

Darwin's idea that complex lifeforms evolved from simpler ones is supported by the fact that fossils of simpler organisms are found in older rocks, whilst newer rocks contain fossils of more complex lifeforms.

I have never found a fossil myself, but I have seen and held one, and felt a real sense of awe in the fact that it was a physical trace of something that was alive millions of years ago. We now have fossils of thousands of different organisms, and they can be arranged in chronological order, from oldest to most recently formed, producing a 'fossil record'. One way of disproving evolution would be to find fossils that showed more complex organisms existed before any less complex ones. By this point you shouldn't be surprised to know that this hasn't been done.

Scientists continue to search for new fossils because they are such a rich source of information about ancient lifeforms, providing a literal snapshot of the past. The remains of some organisms have fossilised better than others, and we have particularly good quality ones that clearly show the evolution of horses and humans. You may have heard of the phrase 'missing link', describing the absence of fossils showing every possible intermediate or transitional organism between humans and our primate ancestors. This is sometimes put forward by people who don't believe, or don't want to accept, the theory of evolution by natural selection, as evidence that the theory is wrong.

However, this is yet another misconception, not least because fossilisation requires specific conditions which make it a relatively rare occurrence, so that most organisms (especially soft-bodied ones) do not become fossils. Furthermore, over the long history of the Earth, many fossils will have been destroyed by events such as earthquakes, volcano eruptions and the natural erosion that continuously takes place.

Fossils of long-dead organisms are not the only evidence we have for evolution. Similarities between living creatures also support the theory that different species are related and descended from common ancestors. The anatomy of all birds is very similar – they all have feathers, wings and beaks. The same is true of insects, which all have segmented bodies and six legs. Many vertebrate species, as diverse as humans, whales and birds, all have a pentadactyl limb, one with five digits. There is no *need* for all these animals to have this similarity, but it can be explained if they all evolved from the common ancestor.

We also see evidence of this when looking at the development of embryos in different species.

For example, all vertebrate embryos have gill slits and tails at some point in their development, which remain as the embryo develops for animals such as fish which need them, but disappear for those, like humans, which don't. This leads us to something else worth sharing with anyone who asks the question, 'Are fish animals?': the science that tells us that fish are animals also tells us that, in the long distant past, one of our ancestors must have been a fish. We have further evidence of this, and of other aspects of evolution by natural selection, from modern techniques (beyond the scope of school science) that allow us to compare our DNA with that of fish.

If my attempt to summarise the evidence above has not convinced you of the validity of the theory of evolution by natural selection, perhaps *seeing* it happen in real time might. We can do exactly this if we study organisms like fruit flies and bacteria, which have short lifespans and can reproduce quickly. Whilst this may be interesting for scientists, it also poses a problem when it comes to human health. Evolution by natural selection has led to the emergence of illness-causing bacteria that are resistant to many antibiotics.

A simple explanation of how this has happened is that when a particular antibiotic medicine is given to people with bacterial infections, a tiny fraction of the millions and millions of bacteria in these people may randomly have a genetic mutation, which means they are not affected by that particular antibiotic. Since they are not affected, these bacteria have a massive *evolutionary advantage* over the other bacteria in the person – they don't get killed, so can go on to produce lots of offspring which are also resistant to that antibiotic. Because bacteria reproduce quickly, there are soon enough of these bacteria that other people being infected with them becomes more likely.

This is why many doctors are reluctant to use antibiotics unless patients absolutely need them – they know that using an antibiotic drug creates a situation in which natural selection leads to the evolution of antibiotic resistant bacteria. It's somewhat worrying that this might, one day, make antibiotics useless as a treatment for bacterial illness.

The above is not an exhaustive list of all the evidence there is for the theory of evolution by natural selection, but I hope I have done a sufficiently good job of explaining the idea that you feel able to understand it. Scientists, teachers and writers far more capable than me have written books and essays trying to do the same thing. Most would agree with the Russian evolutionary biologist Theodosius Dobzhansky that 'nothing in biology makes sense except in the light of evolution'. It is a theory with exceptional explanatory and predictive power.

However, for some people, there is no degree of clarity in an explanation, or amount of evidence to support it, that would convince them of its validity. It may surprise you to know that I have met more than one science teacher who, despite *understanding* and even *teaching* the theory, does not *believe* it. They are not unusual. Many smart people can explain the principles of evolution by natural selection, but do not accept it as 'real' in the same way they accept gravity as being 'real'.

I've already told you about one reason why this might be true – it can be hard to let go of our instinctive sense that complex living things must have been 'designed', despite the fact that evolution by natural selection can *explain* perfectly well the apparent 'design' we see in nature without the need for a supernatural designer (one that is somehow 'above' or 'beyond' the natural world).

I suspect the difficulty many people have with evolution has something to do with the way the theory affects us on an *emotional* level, in a way that learning about gravity or atoms doesn't. Even those of us without religious beliefs can have a sense that we humans are *special*, somehow separate from the rest of nature. We might accept that bacteria arose due to evolution, but it feels wrong to accept that *we*, with our thoughts and feelings and dreams, are the result of the same *mindless* physical process.

Darwin himself understood that people reading his ideas would feel this way, that they would be troubled by its implications for what it means to be alive, what it means to be a human. So, in the very last sentence of his book *On the Origin of Species*, he tells us that:

> *There is grandeur in this view of life, with its several powers, having been originally breathed into a few forms or into one; and that, whilst this planet has gone cycling on according to the fixed law of gravity, from so simple a beginning endless forms most beautiful and most wonderful have been, and are being, evolved.*

I am a little ashamed to admit that I failed to appreciate the *grandeur* Darwin wrote about until long after I left school. I have to confess that I feel I was let down by my biology teachers in this respect. As a teenager, I didn't find biology particularly interesting, and cannot even remember being taught about evolution.

In contrast, I loved my physics lessons, and was left with the impression that it was the only science that dealt with the big ideas, the fundamental questions about the nature of reality. I was so excited by the little I learned at school about Einstein's relativity and quantum theory that I decided physics was the subject I *had* to study at university.

Unfortunately, actually studying these things did not live up to my (perhaps naive) hopes and expectations and, as you may have gathered, I did not become a physicist. However, my degree in physics paved the way for me to become a science teacher, and one of the many good things that happened as a result is that I finally understood why evolution by natural selection is, by far, the greatest scientific theory anyone has come up with, and arguably one of humanity's greatest achievements in any field.

Chapter 7
What Am I Made of?

'Sugar and Spice and all things nice', according to the traditional nursery rhyme, are what little girls are made of, whilst little boys are apparently made of 'snips, snails and puppy-dog tails'. Some people are said to be made of 'stern stuff' and others will insist that they are 'not made of money'. By this point, I hope it will come as no surprise to you that science tells us we are made of *atoms*, like everything we see around us. But the existence of poetry (including nursery rhymes) and metaphors about what we are made of suggest that we are not *really* 'like everything else around us': that human beings are *special*.

A lot of people believe that what separates us from everything else in the world is that we have a 'soul' or 'spirit'. Some people believe that their pets and other animals are also special in this way. These are ideas that are often associated with religion. Scientists prefer to try to understand *consciousness*, and how *sentience*, the ability to experience thoughts and feelings, can come into existence.

We're probably a long way from understanding these things, and there's no guarantee we will ever have satisfactory scientific answers to all the questions we have about what it means to be a thinking, feeling person. One reason for this is that humans, and living things in general, are the most *complicated* and *complex* things we know of in the natural world – complicated in the sense that we are not simple or easy to understand, and complex in the sense that we are made up of many, many intricately connected parts. This is also why I've left this question until the end of the book – you'll need some of the ideas from previous chapters to best appreciate my answer to this one.

In 'Why Does Ice Cream Melt?', I explained how we can answer this question in terms of the particles that ice cream is made of, and what those particles do when it gets hot. Most students I've taught are able to grasp this model of matter, and accept it without much trouble. It's not too difficult to imagine ice cream being made of particles that hold on tightly to each other when it's cold, and less and less tightly as it gets warmer. It's far more difficult to imagine, or accept, that the same kind of particles we find in ice cream, or chairs, or teaspoons, can be put together in such a way that they make *us*.

In another earlier chapter, I tried to answer the question 'What are stars?', and I hope you'll remember that they are mostly very large collections of protons, neutrons and electrons bumping into each other, and releasing light and heat when they do so. The precise details of the mass of a star, its size, its temperature, and so on, can be expressed in numbers and put into mathematical formulae so that, for example, we can predict how long it will shine or determine how far away it is from us. Sure, it would be difficult for most of us to do, but the fact that we *can* describe how a star works using mathematics is, to scientists at least, an indication that they are relatively simple objects.

There is really only one *process* going on in a star – nuclear fusion, and it's one that we understand so well that we can reproduce it in machines here on Earth. The difference between a star and even the *simplest* living thing is like the difference between a single (enormous) test-tube with just two elements reacting together, compared to a whole factory full of different reactions going on in every test-tube, beaker, conical

flask and other available container, with Bunsen burn-
ers and extractor fans going full blast. There are many
more processes going on inside the tiny cells of your
body than in a star, and it's this complexity which
means that the science of living things cannot be
expressed in a few equations.

The lack of equations in school biology often leads
students to believe that it is the easiest of the sciences.
Physics is commonly thought of as the most difficult, partly because of the way schools test students'
knowledge and understanding of a subject – getting
high marks in a physics test usually requires students
to be good at maths and problem solving, skills that are
not required to the same extent in school biology.

Another factor that affects school students' per-
ception of the relative difficulty of these subjects is
that physics and chemistry often deal with unfamiliar objects – atoms, molecules, force-fields, catalysts,
and so on – things we don't generally encounter, or
think about in our everyday lives. These two subjects
also require an understanding of abstract ideas such
as energy and momentum, things we can't see or
touch, but which we are told exist. School biology, on
the other hand, deals mainly with *concrete*, everyday
objects like plants, food and our own bodies. We don't

have to stretch our minds too much to imagine these things, or to believe they exist at all.

However, beyond the confines of the school science curriculum, the differences between physics, chemistry and biology are not so clear-cut, largely because doing scientific research in the modern world cannot be compartmentalised neatly in this way – just ask any astrobiologist, biochemist or biophysicist.

We call the study of living things 'biology', but the different names of the sciences do not really matter. They all stem from the same thing: humans' curiosity about the world around us and how it works. One of the greatest mysteries in science is how life started. Scientists are still trying to figure out the exact details, but they are confident that it is very much a result of chemical reactions. Some scientists have even said that 'all life is chemistry', so it is worth remembering at this point that:

Chemical reactions occur when two or more substances are brought into contact with each other and interact to form one or more new substances.

It turns out that the seven MRS GREN processes that characterise living things – movement, respiration, sensitivity, growth, reproduction, excretion and nutrition – can *all* be achieved through chemical reactions. Right now, as you read this, there are millions and millions of them happening in your body. Together, these reactions are what make you 'alive', and they are collectively known as your 'metabolism'. There are a huge number of different reactions that make up the metabolism of any living thing, but school science focuses on only a few, mainly related to eating and breathing.

Perhaps because it is so obviously important to our health and survival, children often spend quite a lot of time learning about *digestion*. This is the process by which the chemicals we take into our bodies, in the form of food and drink, are broken down into molecules that can enter our bloodstream. For this to happen, there are a variety of physical and chemical processes which need to take place. You can see this for yourself by doing a classic 'practical' at home:

> *Take a piece of bread and chew it for about three minutes. If you manage to chew it for this long without swallowing, you should notice that the mush in your mouth starts to taste distinctly sweet. This is because chewing the bread mechanically breaks it up into small pieces, and mixes it with saliva produced by the salivary glands.*

Saliva contains various substances known as
'enzymes', which react with a chemical in the
bread, starch, and turn it into sugars made up of
smaller molecules, such as maltose and glucose,
that taste sweet to us.

There are similar mechanical and chemical processes
that take place in the different parts of our *digestive*
system which, as well as the mouth and salivary glands,
includes the oesophagus, stomach, pancreas, liver, gall
bladder, intestines and anus. The molecules that enter
our bloodstream as a result of digestion then take part
in chemical reactions, which do everything from building new cells to releasing energy. It often astonishes
people to realise that the phrase 'you are what you eat'
has a literal meaning in the sense that the atoms we are
made from come from the food we eat.

We also take chemicals into our body when we
breathe in. The air around us contains a mixture of
elements and compounds, including water, nitrogen,
carbon dioxide and oxygen. Just as the chemicals in the
food we ingest are processed by the digestive system,
the chemicals in the air we take in are processed by the
respiratory system to extract oxygen, which is essential
for a process that happens in nearly every single one of

our cells: *respiration.* This is arguably the most important process that takes place in living things, because it provides the energy needed for all the other living processes to happen.

Students often confuse breathing and respiration as being the same thing, because the two are connected in humans and other animals. Breathing, or *ventilation* as it is sometimes known, is not a chemical reaction, it is the mechanical process of moving air into and out of your lungs. Respiration is a general term for several chemical reactions in living cells that break down chemicals from food to release energy. In school, students are usually required to learn about two forms: *aerobic* respiration, which needs oxygen, and *anaerobic* respiration, which does not. Aerobic respiration takes place in most of our cells all the time. It can be summarised as:

glucose + oxygen → carbon dioxide + water

We breathe out the carbon dioxide and some of the water produced in this reaction. Again, you can see this for yourself by blowing onto a cold mirror or window – the 'fog' that forms is made of tiny droplets of condensed water from your breath. It's not so straightforward to show the presence of carbon dioxide, but students in school will often be given the opportunity to blow through a straw into a test tube of 'limewater' (a solution of calcium hydroxide in water) and watch as it turns cloudy as a result of a reaction with CO_2.

It's easy to understand that we put on weight if we eat too much food because we are putting more stuff into our bodies, but most people are surprised to learn that the way we *lose* weight is by breathing out molecules of carbon dioxide. Sadly for many of us, this does not mean we can lose weight simply by breathing out more quickly – you would lose only a tiny amount more this way, and be at risk of fainting from hyperventilation. The amount of carbon dioxide we exhale depends on how quickly we are producing carbon dioxide in our cells through respiration, and this only increases significantly when we are doing physical activity.

Another important feature of respiration is that it is an *exothermic* chemical reaction, one which *releases* energy, so it is sometimes written as:

glucose + oxygen → carbon dioxide + water + ENERGY

Sharp-eyed students often notice that this looks like the equation for photosynthesis, but backwards. In that reaction, which we looked at briefly in the previous chapter, carbon dioxide and water are combined to make glucose and oxygen, but this can only happen with the *input* of energy (in the form of light), making it an *endothermic* reaction.

It would be impractical but, if you wanted to, you could have a heater in your home that made use of the same reaction as aerobic respiration. You would need to take some glucose (a type of sugar) and set it on fire. Once on fire, it would react with the oxygen in the air and produce carbon dioxide and water whilst heating up the surroundings. As long as you kept adding glucose to the fire and made sure it was well supplied with fresh air, you could continue to enjoy the heat it produces. If you ran out of glucose or stopped fresh air getting into the room, the fire would go out and the room would get colder.

Similarly, if you stopped eating (to get glucose) and breathing (to take in oxygen), respiration could not happen in your cells and you would die. Living things behave in such a way that their supplies of the chemicals they need for respiration, and all the other processes that keep them alive, do not run out as long as they live. In other words, the chemical reactions in a living thing are 'self-sustaining'. This is such an important property of living things that astrobiologists at NASA, scientists who are looking for life beyond our planet, *define* life as:

> *a self-sustaining chemical system capable of Darwinian evolution.*

This idea is also one that many scientists think can explain the origin of life on Earth. The evidence suggests that the first 'lifeform' on Earth was the result of a chemical reaction that became self-sustaining and developed into something that was able to undergo evolution by natural selection.

Elementary Life

It might seem impossible that non-living, non-thinking atoms could somehow come together, *by themselves*, so that billions of years later, things as wondrous as a flower, a butterfly, or you could exist. But the theory of evolution by natural selection tells us that this is entirely possible, given enough *time*. As I explained in the previous chapter, one of the reasons why it might be difficult for many of us to accept this astonishing idea is that it took *billions* of years for this to happen, a timescale our minds cannot easily imagine, if at all.

Once the first cell came into existence, it took another two billion years for multi-cellular life-forms to emerge and around a billion years more for humans to appear. Today, there are millions of different species on Earth, but *all* are made up from a relatively small selection of the hundred or so naturally occurring elements (which, I hope you remember, are substances consisting of just one type of atom). This makes sense if we accept that all life started from the same original set of chemical reactions, which somehow became self-sustaining.

It turns out that a valid answer to the question 'What am I made of?' could be 'mostly oxygen, carbon, hydrogen and nitrogen', because these four elements make up about 96 per cent of all the atoms in any living thing. The remaining 4 per cent are mostly calcium, phosphorus and sulfur, with tiny amounts of other elements including sodium and iron. In our bodies, the atoms of these elements don't usually exist alone, but are found joined together with other atoms as part of one of many different molecules.

The most abundant molecule in living things is water (H_2O), so another technically correct answer to the question at the start of this chapter is 'mostly water', because more than half of a human body is made of water.

Some organisms, like jellyfish, can be more than 90 per cent water. Although DNA is perhaps the most famous chemical associated with living things (and we will look at why that is later in this chapter), water is in many ways more important. One of the special properties that liquid water has is that lots of other substances can dissolve in it, so that it can 'hold' lots of different atoms and molecules in such a way that they can react with each other. The other chemicals needed for life would not be able to do their 'jobs' without being in water.

Life on Earth started in water, over three and a half billion years ago, and only moved onto land less than half a billion years ago. Although many organisms now live outside of the water, nearly all the important processes that keep them alive still happen *inside* the water that is inside every single one of their cells. The role of water, in life as we know it, is so fundamental that scientists looking for extra-terrestrial life usually focus on planets with water on them. The Nobel Prize–winning biochemist Albert Szent-Györgyi summed this all up as:

Life was born in water and is carrying on in water. Water is life's mater and matrix, mother and medium. There is no life without water. Life could leave the ocean when it learned to grow a skin, a

> *bag in which to take the water with it. We are still*
> *living in water, having the water now inside.*

Water is the key component of cytoplasm, a jelly-like substance that forms the bulk of most cells. Other parts of a cell are made from different molecules, including fats, proteins, sugars and DNA. These molecules which make up the structure of a cell, and indeed most of the molecules involved in living processes, have long strings of carbon atoms in them, so scientists sometimes refer to all life on Earth as 'carbon-based'. There are so many carbon compounds that the study of them is a scientific field of its own, known as 'organic chemistry'. This name was originally given to the study of chemicals that were derived from living things.

Before the 1800s, many scientists believed in 'vitalism', from *vita*, the Latin word for 'life'. This was the idea that plants and animals contain some sort of distinctive quality, a 'force' or 'spirit'. According to this belief, 'organic' chemicals could only be produced by living things. Vitalism was eventually proved wrong by many scientists, including the German chemist Friedrich Wöhler, who showed that urea, a substance found in urine, could be made using chemicals that were definitely not from living things. Today, the molecules which exist inside organisms, and the chemical reactions they take part in, continue to be studied by scientists who want to deepen our understanding of the processes that, together, make something alive.

Cells

Whilst atoms and molecules are the building blocks which make up all living things, no single chemical can be described as being 'alive'. The *simplest* thing that meets all the MRS GREN requirements to be described in this way is a *cell*. Unlike atoms, we don't have to imagine what cells are like, we can *see* them, if we have access to a sufficiently powerful microscope. As I've already argued in 'What Is the Smallest Thing?', I think a half-decent science education should include the opportunity for students to use a microscope and to look at cells for themselves. If possible, students should look at more than one type of cell to appreciate that they come in different shapes and sizes, but have many of the same features.

The most obvious feature of a cell is that it has a *membrane*, the thin barrier that keeps its insides separate from the outside, a bit like the skin of a balloon keeps the air inside separate from the air outside. Cell membranes are *semi-permeable*, meaning that some molecules can move in and out of the cell, whilst others are blocked from entering or leaving. This is important, because the chemicals used and produced by metabolic reactions need to be able to enter and leave the cell. One of the things students are generally asked to notice when looking at plant cells is that they *all* have a 'wall', made from cellulose, which surrounds the membrane. No animal cell has such a wall, making this one way of determining if something is an animal or a plant.

Inside the membrane of all cells is the *cytoplasm* which, as I mentioned above, is a jelly-like substance consisting mostly of water. This contains nutrients and other chemicals dissolved in it, as well as structures called *organelles* (literally 'little organs'), which carry out specific processes needed to keep the organism alive. Most school microscopes will be powerful enough for students to see *chloroplasts*, the organelle in plant cells where photosynthesis takes place. Respiration in both plant and animal cells takes place in much smaller organelles called *mitochondria*, which cannot be seen with a school microscope.

Mitochondria are often described as the 'power plants' of a cell, because they make a chemical called *adenosine triphosphate* (ATP), which stores the energy from respiration and makes it available for the other processes that happen in a cell. The biggest, most easily seen organelle in most cells is the *nucleus*, appearing as a small, dark blob inside the cytoplasm. This is often referred to as the 'control centre' of the cell, a phrase I don't particularly like because it can lead children to have the misconception that the nucleus is somehow the 'brain' of a cell. However, the nucleus, as indicated by its name, does play a central role in the functioning of a cell, because it contains DNA.

Unfortunately, school microscopes are not powerful enough for students to see individual bacteria. These unicellular (single-celled) organisms do not have a nucleus and are classified as *prokaryotes*, meaning 'before nucleus' from the Greek words for 'before' and 'kernel'. Cells with nuclei are known as *eukaryotic*, from the Greek for 'after kernel'. The first cells that existed were prokaryotic and for a long time it was not clear how eukaryotes could have evolved from them. The American biologist Lynn Margulis helped to solve this problem by providing evidence that eukaryotic cells could have been formed through the *merger* of two or more different types of prokaryotic cells.

This process, called *endosymbiosis* (from Greek words meaning 'inside' and 'living together'), was initially controversial because it contradicted the idea that evolution usually occurs in tiny steps (it *usually* does, but not always), but is now widely accepted as the process that gave rise to mitochondria and chloroplasts in cells, and so the most likely way in which the first eukaryote came about. Similarly, scientists think the first *multicellular* organisms (living things made of more than one cell), probably came into existence when two single cells stuck together and became an organism that was better equipped to survive in its particular environment.

Since then, multicellular life-forms have evolved so that different cell types, like nerve cells and muscle cells in humans, do different jobs to keep the organism alive. Similar cells are joined together to make *tissues*, like the epidermis of our skin or the marrow of our bones. Tissues form *organs*, like our hearts, livers and brains. Organs are usually part of organ *systems* such as our cardiovascular, digestive and nervous systems. Finally, at the top level of this hierarchy of complexity is the whole *organism*. So a more complete answer to the question 'What am I made of?' is:

You are made of organ systems, which are made of organs, which are made of tissues, which are made of cells, which are made of molecules, which are made of atoms.

Atoms were *predicted* to exist by scientists and philosophers long before we had any evidence for them, but nobody had any idea that cells, as we know them, existed until we could look at living, or once-living, things under a microscope. Although Robert Hooke used the term 'cell' as early as 1665 to describe the structures he saw when looking at some cork under a microscope, there is no evidence he appreciated how important cells would be to understanding living things. Over the next two hundred years or so, many other scientists used microscopes to look at lots of different cells, and shared their findings by publishing books and articles about their work.

Eventually, the fundamental role that cells played in living things became clear, and by the 1860s scientists had developed a *cell theory of life* based on these principles:

- All living things are made of one or more cells.
- The cell is the most basic unit of life.
- All cells come from pre-existing cells.

Matthias Schlieden and Theodor Schwann are usually acknowledged as being the first to write down the first two principles, while Rudolf Virchow is often given credit for coming up with the third. Some historians of science believe that Virchow may have plagiarised the idea from his Jewish colleague Robert Remak, who had already found evidence in the early 1840s that new cells are produced when existing ones split in two.

It is important to note that Remak was Jewish, because this was literally why, despite considerable accomplishments as a scientist, he was not allowed to be a university professor in parts of Poland that were under the rule of antisemitic Prussians (Germans). Prejudice against Jewish people may also be why Remak's work was not more widely appreciated elsewhere. (Antisemitism had far worse consequences for one of Remak's sons, who was killed by the Nazis in Auschwitz.) The development of cell theory, based on research done by lots of people, is a good example of how, despite the impression that is often given in school science lessons and textbooks, major advances in science are hardly ever the result of just one scientist's work.

Bacteria

One of the classic biology practicals in school science lessons involves growing bacteria collected from different surfaces in the classroom or perhaps even from around the whole school. Students are supplied with some cotton swabs and several shallow round glass

Petri dishes with a thin, cloudy disk of 'nutrient agar jelly' stuck inside. The swabs are rubbed on the surface being tested before being rubbed onto the agar jelly in a dish. The dishes are then labelled and left in a dark, warm cupboard or an 'incubator', depending on how fancy the school is. After a week or so, there are usually some impressive-looking, slightly disgusting, yellow or greenish or even red-coloured blobs which have appeared on at least some of the dishes.

These are 'colonies' of bacteria and, in my experience, students are normally surprised to find that the biggest ones are not from the toilet bowl or the bottom of their shoes, but places such as the classroom door handle or computer keyboards. This can be explained by the fact that a school toilet bowl is washed every time you flush it and regularly disinfected, but a door handle or computer keyboard is cleaned less often, and has been touched by countless people who, even after the world has gone through a major pandemic, probably don't wash their hands after going to the toilet or sneezing or coughing into them.

This activity is usually done to show students that *microbes* (microscopic organisms) are present all around them, and perhaps to challenge their ideas of which surfaces harbour the most bacteria. It also shows how quickly bacteria can reproduce – the reason the colonies of bacteria can be seen is because they are made of billions and billions of bacteria, which are descended from the original bacteria collected on the swab.

If the temperature and other conditions are just right, and there is plenty of food available (like in a nice warm Petri dish filled with nutrient agar jelly), the number of bacteria in a sample can double once every twenty minutes or so. This means that a *single* bacterium can become eight bacteria in an hour, 64 in two hours, 512 in three hours, over *two million* in seven hours and over a *billion* (a thousand million) in just under ten hours. This doubling and doubling of the bacteria is an example of *exponential growth* – their numbers *increase at an increasing rate*, as long as they don't run out of food and other necessities.

The bacteria, like other prokaryotes, reproduce through a process called *binary fission*, which literally means splitting in two. During this process, various chemical reactions take place inside the cell, resulting in copies being made of all its key internal components. The cell then becomes elongated so that each end of it contains all the things the original cell did, and then it splits. The resulting two cells may be ever so slightly different physically, but they are *clones* of the original cell, and of each other, meaning that they have the same DNA and are *genetically* identical.

Another practical that I think everyone ought to do at school or at home, is grow a plant from a seed. If you've never done this, I highly recommend it – there is something magical about planting a seed in some soil and watching a plant emerge from it. You don't even need soil: some seeds, like cress or pea, will sprout if you simply put them on some damp tissue paper. If you do this in a glass with the seed pressed up against the side, you can watch the roots as well as the shoot emerge.

I've tried my best to keep hyperbole to a minimum in this book, but I cannot resist telling you that I think seeds are *ridiculously, extraordinarily* amazing. I know I'm not the only person who has marvelled at the fact that a seed can grow into a plant or a tree, and wondered *how* it's possible for this to happen. Obviously an acorn doesn't *contain* an oak tree. It doesn't *expand* into a tree. Instead, the acorn somehow *makes* the tree, like some sort of tiny 3D printer.

The seed starts off containing the materials it needs to start 'printing' the oak tree. It just needs to produce the roots and top growth, and then it can get more 'ink' by the process of photosynthesis, with its leaves taking in atoms from the air while its roots take in atoms from water in the ground. The 'printing' is powered by respiration and the tree *grows*, like the bacterial colonies described above, by *cell division*. The process is not quite the same though, because bacterial cells only produce copies of themselves, but some of the cells in an acorn somehow turn into the roots, bark, leaves and other parts of the oak tree.

Acorns, like all other plant seeds, contain *stem cells*, which have the ability to *differentiate* to form *specialised* cells. And it's not only trees and plants that grow like this – *all* multicellular organisms, including humans, start off simply as one cell which splits over and over again, with the cells *differentiating* to form tissues and organs and all the other parts that together make the whole lifeform.

All this raises lots of questions, some of which scientists are still struggling to answer. The mystery of how cells 'know' how to turn into different kinds of cells is linked to something that baffled Charles Darwin and other scientists well into the 1900s – how do organisms inherit traits from their parents? Or to put the question of *inheritance* in the child-like manner of the others in this book, *why do I have my mum's nose?*

The answer lies in the most famous three letters in science: *DNA*.

DNA

Unlike H_2O, which tells us that there are two hydrogen atoms and one oxygen atom in a molecule of water, DNA is not a chemical formula, but an abbreviation of the words *deoxyribonucleic acid*. This substance can be found inside the cells of *every* living thing, usually as incredibly long molecules known as 'polymers'. It's perhaps the only chemical substance for which many non-scientists could draw at least a rough picture of its structure, showing its atoms arranged into something that looks a bit like a twisted ladder.

There can't be many school biology departments in the world that don't have a poster or model depicting it, but even if you've never set foot in a school or opened a biology textbook, you've probably seen a representation of this famous 'double helix'. Like the solar system model of the atom, the periodic table and the equation $E = mc^2$, the structure of DNA has become an iconic image of science, embedded in our common consciousness as the universal symbol for our scientific understanding of life.

The images of DNA with which we are familiar are not actually what it 'looks' like because, as I explained in 'What Is the Smallest Thing?', atoms are too small for us to see with our eyes, even with the most powerful light microscope. The structure of DNA had to be *worked out* using a technique known as 'X-ray diffraction', which involves firing X-rays through a sample of a chemical, and then using mathematics to interpret the resulting image. The details of exactly how this works are related to the fact that, as I've mentioned before, mathematics can be used to describe and predict the behaviour of simple things such as particles and X-rays.

The story of how the double helix was discovered from an X-ray image has been covered in many books and documentaries, and is one of the most well-known tales from the history of science. Its fame is not solely because of its scientific significance, but also because it highlights sexism in science and how the role women have played in scientific discoveries throughout history has often been ignored or deliberately misrepresented.

In this case, three men, Francis Crick, James Watson and Maurice Wilkins, were awarded the Nobel Prize for the discovery of the structure of DNA, while the work of Rosalind Franklin, which was indubitably necessary for this breakthrough, was not given appropriate recognition at the time. Franklin's work, and accounts from people who knew her, show that she was clearly an exceptional scientist and a remarkable woman in general, who was often not given the respect or status she deserved from her male peers due to their sexist attitudes. A lot has been done since then to acknowledge Franklin's contribution, and recognise her brilliance, but she died at the age of thirty-seven and did not live long enough to see this, nor to be awarded a Nobel Prize.

The reason why the DNA double helix has become so famous is that the structure can explain how this particular chemical can do two things which are absolutely necessary for life:

> *DNA encodes the information necessary for an organism to develop from a single cell and for all the other processes it needs to live, and it provides a way for that information to be copied and passed on to new cells and offspring.*

To understand how DNA works, imagine two separate legs of a ladder that both have a series of half-rungs sticking out of them. These half-rungs have to be slotted together to make the ladder whole. Now imagine that each half-rung has a letter painted on it, either an A, C, G or T, and that half-rungs with the letter A can only slot together with a half-rung marked T, and those marked C only fit with those marked with a G.

This means that if you have only one side of such a ladder, you will always be able to construct the other half, because you would know what letters would *have* to be on the missing half. Furthermore, if you wanted to make copies of the original ladder, you would need only one leg to do so – because that one leg contains all the information you need to assemble the other.

This is a simple model for how polymers of DNA reproduce – they split in half and each half then gains a new half, making two identical polymers in the place of one. In the case of DNA, the 'half-rungs' with the letters on them are chemicals known as 'bases' (short for nitrogenous base), which have the names adenine (A), cytosine (C), guanine (G) and thymine (T). The order in which these four bases occur in a molecule of DNA is known as a DNA *sequence*.

As well as providing a mechanism for reproducing DNA, these letters also provide a way of encoding instructions for the chemical reactions that need to take place in a cell – just as you can extract information from the sequences of letters you are reading now, a cell gets the information it needs from the sequence of bases along DNA molecules.

More specifically, the instructions encoded in DNA determine which proteins are made in cells, and these proteins are the chemicals that control all the reactions which, together, allow an organism to grow and live. I must stress again that this is a very simple model of how all this works – chemicals do not 'read' other chemicals in the same way that you are 'reading' this, but I hope you can see how, at least in principle, information can be stored and duplicated by DNA.

In humans and other eukaryotes, DNA is found mainly in the nuclei of our cells, in structures called *chromosomes*. These are single, very long, extremely coiled-up molecules of DNA. Humans have 46 of these in each of our cell nuclei, arranged in 23 pairs, with one of each pair coming from each of our parents. Other organisms can have as few as two chromosomes or well over a thousand; the number of chromosomes does not appear to be related to the complexity of an organism.

Each chromosome, or long molecule of DNA, is made up of hundreds, or even thousands, of small sections called *genes*. Each pair of chromosomes contains the same genes in the same positions. The order of the bases A, G, C and T in a gene provide a code for making a particular protein that will be responsible for specific chemical reactions. These reactions then lead to *everything* that cells do, from making copies of themselves, to forming tissues, organs, organ systems, and ultimately, whole organisms. To go back to our 3D printer analogy, genes are like the instructions a computer sends to such a printer, where the DNA sequence of the bases A, G, C and T is analogous to the ones and zeroes that make up the binary code used by computers.

It's not uncommon to hear the phrase, 'It's in their genes', when parents are proud of their children for doing something. Most people are also familiar with the idea that there are genes that can be passed on from parents which make it more likely their children have a disease such as sickle cell anaemia or cystic fibrosis. In other words, even people who don't have much scientific knowledge are aware that genes are *units of heredity*.

Using this idea, we can answer the question 'Why do I have my mum's nose?' with 'because you have her genes'. The idea for a 'unit of heredity' was around long before the discovery of DNA. One of the scientists usually credited with coming up with the idea is Gregor Mendel, a monk who carried out experiments with pea plants to determine how inheritance worked. Accounts of Mendel's work often emphasise the impressive fact that he grew thousands of plants and he was meticulously methodical in his work. What is mentioned less often is that hardly anyone took notice of Mendel's work at the time, and it was not until other scientists independently carried out similar work, decades later, that his work was rediscovered.

You've probably gathered by now that I'm not a massive fan of the classic stories of 'scientific discovery' that are often part of a school science education. There are two reasons why I try to provide more accurate accounts, rather than reinforcing the myth of the lone genius. First, is to clarify that science is more often than not a communal endeavour, one carried out by many people working together, both formally and informally. Second, and perhaps more importantly for school students, these stories can convey the elitist idea that science is something done only by brilliant individuals, and not something that pretty much anyone can aspire to.

None of this is to diminish Mendel, who was undeniably a good scientist, and was one of the first to take a mathematical approach to a problem in biology. By recording how various features, like the colour of flowers, developed in his pea plants, Mendel was able to show that there must be some be some 'unit' or 'particle' that was passed on from parent to offspring which determined how these characteristics were inherited. More than this, he was able to show that it was possible to *predict* what proportion of plants would inherit particular traits, because he had a *mathematical* model for how these 'units of heredity' were passed on, based on the data he had collected. Although he didn't call them by the same name, later scientists agreed that what Mendel had identified was what we now call *genes*.

Genes

One of the classic examples of genetic inheritance taught in schools is how eye colour is determined. The fact that people have different coloured eyes tells us that there must be different forms of the gene for eye colour. These different forms of a gene for a particular characteristic are called *alleles*. They are the same 'section' of a chromosome, but with the bases A, G, C and T occurring in a slightly different order. Children inherit one allele from each parent, so you can have the two different alleles for the same gene in your chromosomes. The particular combination of alleles you have in your chromosomes for a trait is called your *genotype*.

My wife has blue eyes and I have brown eyes. We have passed on different alleles for eye colour to our daughters, who both have brown eyes. According to

Mendel's ideas, this is because, despite inheriting an allele from their mother that would produce blue eyes, the allele they have inherited from me, for brown eyes, is *dominant*. The allele for eye colour they have inherited from their mother is *recessive*. The dominant allele of a pair of genes will mask the presence of the other one, so even though both of your parents made an equal genetic contribution, some traits will look more like one parent than the other.

The idea that one gene is responsible for one characteristic, such as eye colour, has been useful for scientists as the science of genetics has developed but today we know that things are a bit more complicated. Even something as apparently straightforward as eye colour cannot always be explained in terms of the simple Mendelian genetics that children are taught in school. Today's scientists know that eye colour, like most other physical traits, is ultimately a result of the combined action of multiple genes along with many external factors, including things such as our diet and environment. For example, my youngest daughter was born with blue eyes, which slowly became darker until they were brown like mine and her sister's. This is not an uncommon phenomenon. My daughter's eyes eventually became brown because the cells in her iris, the

coloured part of her eye, gradually produced more mel-
anin, the chemical responsible for brown skin. The precise rate at which this happened, and the exact shade of
brown, will have been the result of the actions of over
60 genes now known to be involved in eye colour, and
perhaps also due to the effects of the environment in
which she has grown up.

It is usually an over-simplification to talk about *a*
gene 'for' a particular trait; genes provide instructions
for certain chemical processes to take place in our cells,
but the outcomes of those reactions are also influenced
by what else is going on in and around the cells. This
is why even identical twins, who have exactly the same
DNA, are different.

Humans receive one set of chromosomes from each
of our parents because we cannot reproduce as individuals – there needs to be two of us involved in the process.
The way we produce offspring, where two parents are
needed, falls into a category known as *sexual* reproduction. Most organisms reproduce in this way but there
are some, like bacteria, starfish and some plants, which
can self-replicate through a process known as *asexual*
reproduction.

This produces genetically identical offspring because
it comes about from a process of splitting cells called

mitosis, which produces identical copies (clones) of the parent cells. Sexual reproduction produces offspring that are genetically different from their parents, because genes are mixed or 'shuffled'. For sexual reproduction to take place, each parent produces special 'sex cells' known as *gametes*. These are produced by a process known as *meiosis* and are different from other cells, because they have only half the number of chromosomes in an ordinary cell.

The new cells produced as a result of meiosis get a mixture of chromosomes from the pairs of chromosomes in the original cell, making the gamete genetically different from the original cell. This genetic variation is one reason why offspring produced in sexual reproduction are different from their parents. This genetic variation is crucial for evolution.

In many organisms, there are two distinct types of gamete, male and female. In humans, the male gametes are sperm, and the female ones are eggs, each of which contain 23 chromosomes. For reproduction to happen, male and female gametes need to meet so that their nuclei join together, a process known as *fertilisation*. The cell formed by this fusion is known as a *zygote* and it contains the chromosomes from both parents. The zygote then divides by mitosis to form an embryo and eventually the whole new organism.

The word 'sex' has different meanings depending on the context in which it used. Many children associate the word with sexual intercourse, and so have a tendency to giggle whenever a science teacher uses it in class. In biology, 'sex' is the trait an organism has that determines whether it produces a male or female gamete. Sex is often confused with gender, but it's important to know the difference, because our beliefs about them can and do cause a lot of suffering.

According to the World Health Organisation website at the time of writing:

> *Gender refers to the characteristics of women, men, girls and boys that are socially constructed. This includes norms, behaviours and roles associated with being a woman, man, girl or boy, as well as relationships with each other. As a social construct, gender varies from society to society and can change over time.*
>
> *Gender is hierarchical and produces inequalities that intersect with other social and economic inequalities . . . Gender inequality and discrimination faced by women and girls puts their health and well-being at risk.*

The sex of a person is largely determined by a pair of chromosomes known as the *sex chromosomes*. In a female, these are a pair that contain the same genes in the same position, and are called *X chromosomes*. Females are said to have the *genotype* XX. In a male, this particular pair of chromosomes are distinctly different. A male has only one X chromosome and the other one, which is distinctly smaller, is known as a *Y chromosome*. Males have the genotype XY. Again, it's important to remember that these genes play an important role in sex determination, but the *phenotype* of 'being male' is a result of the interaction between the genotype and other factors including the environment in which the zygote develops into an embryo, and so on.

Humans have about 20,000 genes split between our 23 pairs of chromosomes. If you could stretch out all the DNA in just one set of our 23 pairs of chromosomes, it would be roughly 2 metres long and contain a sequence of A, G, C, T, of about 3 billion pairs of these letters. The complete sequence of the DNA in our chromosomes is known as our *genome*.

There is a printed-out version of a human genome at the Wellcome Collection in London, consisting of over a hundred books, each with a thousand pages, printed in a tiny font. This was the information produced by the Human Genome Project in 2003, one of the biggest, most expensive scientific research projects ever carried out. It was an immense, international collaboration that cost over a billion pounds and took more than ten years. Today, you can get your own genome 'sequenced' in a few days for less than a thousand pounds.

Genes make up less than 2 per cent of the DNA in our genome. The rest is sometimes known as 'junk DNA' because, until relatively recently, scientists were not sure what, if any, purpose it served. A better term for junk DNA is 'non-coding DNA', because while the genes 'code' for making proteins, other parts of our DNA seem to be involved in different processes, such as preventing genes from being damaged or being changed ('mutated'). Genes can be damaged by things like radiation or, very rarely, they can change randomly.

While it's important that, for life as we know it to exist, genetic information is copied and passed on with tremendous accuracy, it's also necessary for some variations to occur because they provide a way for evolution to happen. The variations in our DNA are what make us different from one another, so it may come as a surprise that these differences account for less than one thousandth of our genome.

In other words, every human's DNA is more than 99.9 per cent the same. Perhaps more surprisingly, our DNA is around 99 per cent the same as that of a chimpanzee, 85 per cent of that of a mouse and 50 per cent of that of a banana. This shouldn't actually come as a surprise because it confirms what the theory of evolution by natural selection tells us: life on Earth started with one cell and every single lifeform is related to every other.

It's one of the greatest accomplishments of science that we can explain the processes that make us alive in terms of the atoms we are made of. But we are more than simply collections of atoms, more than the sum of our parts. Atoms don't ask questions or tell stories, we do. Answering the question 'What am I made of?' allows us to add to the scientific 'creation story' I have described earlier in this book.

The Big Bang Theory provides an evidence-based account of how atoms came into existence, and our understanding of DNA, combined with the theory of evolution by natural selection, develops the narrative further, revealing how atoms somehow turned into *us*. But the story isn't quite complete yet, there are lots of gaps that need to be filled, lots of questions left to answer.

I hope you've enjoyed reading my attempts to answer the seven questions in this book, and perhaps even learned a thing or two. I didn't always find it easy to write, and I sometimes questioned whether I was being foolish in trying to condense the key ideas of a good science education into this short book. I'm not sure whether I've done a particularly good job – that's for you to decide – but what I've tried to do in the pages of this book is what I've been trying to do in all my years as a science teacher and communicator: share my knowledge and love of this subject in a way that reflects the joy and fulfilment it has brought me. Although I never became a scientist myself, the subject has been central to my life and provided me with ways to flourish that I could never have imagined – including getting to write this book.

I didn't always like science in school, so I understand when other people say they didn't like, or even hated the subject themselves. It was only when I had teachers whose passion and enthusiasm for the subject were infectious that I understood that science isn't just a collection of facts to be recalled and regurgitated, but rather an intellectually thrilling, ongoing, collective, and very *human* endeavour. I hope I have done for you, in some small way, what those teachers did for me.

Appendix

How to make a Jelly Baby Wave Machine
You will need:

100 or so jelly babies
50 or so wooden kebab skewers
About 6 metres of 5 cm wide duct tape
Two wooden coat hangers, or ones which are not easily bent out of shape
Two chairs and some books or other heavy items to weigh them down
Scissors

SAFETY: The kebab skewers are very sharp, so be very careful when attaching the jelly babies. Also, make sure you don't stick the tape anywhere it might leave a permanent mark.

Method:

1. Decide where you're going to build your wave machine and carefully unroll about 3 metres of duct tape on the floor (or on a dining room table if you have one long enough) so that it is *sticky side up*.

2. Starting about 30 centimetres from one end of the tape, place kebab sticks along the tape at about 5 centimetre intervals, leaving about 30 centimetres at the other end of the tape free.

3. Starting about 5 centimetres from the first kebab stick at one end, carefully run another length of tape, sticky side down, along the top of the first tape, so that you end up with the kebab skewers 'sandwiched' between the two layers of tape. Press down firmly on the upper tape as you do this so that the sticks are stuck in place. Make sure to leave 25 centimetres or so of sticky tape left uncovered at the other end.

4. You should have two bits of sticky tape left exposed at each end of your contraption at this point. Place the bar of a coat hanger in the middle of one of these sticky sections of tape and fold the tape over so that the hanger is firmly attached. The tape and the hanger bar should make a 'T' shape. Repeat this with the other end of the tape.

5. Hang each coat hanger from the back of a chair. Move the chairs apart until the tape is in tension and hangs in as a straight line as possible. Put books or something else heavy on the chairs to stop them toppling over. You may also find it helpful to tape the hooks of the hangers to the chairs so that they do not slide about.

6. Carefully impale a jelly baby onto each end of all the kebab sticks.

7. The tape will probably have become slack with the additional weight of the jelly babies, so adjust the chairs to ensure it is as taut as you can make it.

8. Your jelly baby wave machine is ready! Slightly lift up one of the jelly babies at one the end of the tape and let it go and watch what happens.

9. Try gently jiggling the jelly baby up and down.

10. To see how changing the medium affects the wave, remove all the jelly babies from half the length of tape.

You can see a video of how to build and use a jelly baby wave machine on my website at https://alomshaha.com/portfolio/jelly-baby-wave-machine/

How to make a Pinhole Camera

You will need:

1 Pringles or similar crisp tube (wiped out and made clean)

Sellotape

Duct tape or aluminium (tin) foil

Scissors or a bread knife

12 cm by 12 cm piece of greaseproof paper (preferably white, not brown), or tracing paper

A pen

Drawing pin or safety pin

Ruler

Magnifying lens or pair of spectacles (optional)

Method:

1. Use the drawing pin to make a hole in the middle of the unopened end of the crisp tube.

2. Use the ruler and pen to draw a line around the tube which is 5 cm from unopened end of the tube.

3. Use the scissors or bread knife to carefully cut the tube along this line so that you are left with two tubes, one short, one longer.

4. Hold the square of greaseproof or tracing paper over the open end of the small tube and put the

plastic lid over it so that it clamps the paper in place, completely covering the hole.

5. Fold down the edges of the paper sticking out from under the lid, and use the sellotape to stick these edges onto the tube so that the paper is taut under the lid.

6. Remove the lid.

7. Use the sellotape to rejoin the two tubes where you cut them, so that the paper screen is now inside the crisp tube.

8. Wind duct tape around the whole length of the tube. If you don't have duct tape, you can wrap a sheet of tin foil around the whole tube and use sellotape to secure it firmly in place. This is so that no light can get in through the sides the tube.

9. Look through the open end of the tube while pointing the end with the pinhole out of a window. Cupping your hands around the tube when it's held against your eye will help keep out stray light, and you are more likely to see something interesting if it is bright outside and you are in an unlit room.

You can see a video of how to build and use a pinhole camera on my website at https://alom-shaha.com/crisp-tin-camera/

Acknowledgements

One of the things I've tried to convey about science in this book is that it is largely a collaborative endeavour, and the same is true of this book. I really wouldn't and couldn't have written it without the help and support of family, friends and colleagues. I want to thank a lot of people here, but there are two people who I don't think I can ever thank enough – Paul McCrory and David Sang. They were the first people I turned to for advice when I started writing this book. I knew that both of them, with their knowledge, skills and experience, could have written their own, better version of it, and I was utterly shameless in seeking their help whenever I needed it. They both read and re-read every sentence, paragraph and chapter in the book, gently and kindly pointing out my mistakes, inaccuracies and any explanations which just weren't quite right. I can't quite believe that Paul and David didn't end up just sending my emails to their junk folders and blocking my phone number – I really did take advantage of their generosity and I hope they feel that this finished book is somehow worth all the time and effort they put into it. Thank you, gentlemen, I wouldn't bother reading any of the book except this bit if I was you.

I have wanted to write a book like this for a very long time, but the idea of trying to produce a tome about everything covered in school science (I know, I know, I haven't covered 'everything') was always too overwhelming for me to actually sit down and start work on it in earnest. It was only when Harriet Poland from Hodder got in touch with an idea about using school science to answer kids' questions that I finally started seeing how such a book might work. Thank you, Harriet, for the best writing prompt I've ever had.

Harriet introduced me to Izzy Everington, who was supposed to be my editor. She was hugely positive and enthusiastic about the bits of my work she read, and I must apologise to her again for taking so long to deliver the book that she moved on to bigger and better things. Thank you for being so understanding, Izzy.

Rupert Lancaster kindly took on the responsibility for finishing this book with me at Hodder and has reassured me that it does work and that people will read it. Thank you, Rupert, for seeing me through the final stages of the process and making sure this book exists in the form it does.

Others at Hodder who have contributed to the production of this book are Zakirah Alam, Purvi Gadia, Emily Goulding, Claudette Morris, Richard Rosenfeld, Geraldine Beare, Ruchi Bhargava, Kate Brunt and Dominic Gribben. Thank you all for your hard work.

I will always be grateful to my agent Catherine Clarke for convincing me I could even dream of being published. She has been trying to get me to write a book like this for over a decade and I hope this one lives up to her expectations. Thank you again, Catherine, and I look forward to our next project.

The biology chapters of the book were the ones I enjoyed writing the most, perhaps because they were the most interesting and challenging for me. I would not have the confidence to put these chapters out there if they hadn't been checked by some incredibly accomplished biologists – Aoife McLysaght, James Brown, Simon Culver, Joe Wright, Alexander Holmes and Matthew Cobb. I know all of you have busy, demanding lives, and I am incredibly grateful to you for making the time to check (and correct) my physics teacher's knowledge of biology.

Diane Leedham and Kerry Moore kindly read the whole book from cover to cover and helped spot many non-science related mistakes. Thank you both for helping make the book a little better than it otherwise would have been.

Richard Smyth – I hope I don't need to tell you again how much I appreciate your support but for what it's worth, here it is in writing: thank you.

Roma Agrawal, Alex O'Brien and Rebecca Struthers – thank you for making me feel like a real science writer at a time when I couldn't write anything.

Vivian Archer – I don't think anyone has sold more copies of my book than you. Thank you for all you do for writers like me.

I hope the following people will not feel aggrieved that I have bunched them all together, but most of them have been acknowledged in my previous books and will probably be acknowledged again if I write any more – friends who have supported my work for longer than I care to remember and whose brains I pick on a regular basis: Philip Ball, Dean Burnett, Jon Butterworth, Stephen and Anne Curry, Helen Czerski, Luke Donnellan, Declan Fleming, Ronan and Jane McDonald, Sai Pathmanathan, Michael de Podesta, Andrew Ponzen, Elin Roberts, Jonathan Sanderson, Andrea Sella,

Marie-Claire Shanahan, Jon Turney and Alex Weatherall. Apologies to anyone I've missed out, I'll make it up to you by cooking dinner.

Finally, Kate, I won't be (too) soppy and embarrass you in public, but you know I must thank you for everything, especially those two bundles of atoms, Renu and Mina, who make the nothingness of the universe not only bearable but joyful.

Index